MATHEMATICAL LOGIC FOR COMPUTER SCIENCE

Second Edition

WORLD SCIENTIFIC SERIES IN COMPUTER SCIENCE

25: Computer Epistemology — A Treatise on the Feasibility of the Unfeasible or Old Ideas Brewed New (*T Vámos*)

26: Applications of Learning and Planning Methods (Ed. *N G Bourbakis*)

27: Advances in Artificial Intelligence — Applications and Theory (Ed. *J Bezdek*)

28: Introduction to Database and Knowledge-Base Systems (*S Krishna*)

29: Pattern Recognition: Architectures, Algorithms and Applications (Eds. *R Plamondon & H D Cheng*)

30: Character and Handwriting Recognition — Expanding Frontiers (Ed. *P S P Wang*)

31: Software Science and Engineering — Selected Papers from the Kyoto Symposia (Eds. *I Nakata & M Hagiya*)

32: Advances in Machine Vision — Strategies and Applications (Eds. *C Archibald & E Petriu*)

33: Mathematical Foundations of Parallel Computing (*V V Voevodin*)

34: Language Architectures and Programming Environments (Eds. *T Ichikawa & H Tsubotani*)

35: Information-Theoretic Incompleteness (*G J Chaitin*)

36: Advanced Visual Interfaces (Eds. *T Catarci, M Costabile & S Levialdi*)

37: Non-Deterministic Concurrent Logic Programming in PANDORA (*R Bahgat*)

38: Computer Vision: Systems, Theory and Applications (Eds. *A Basu & X Li*)

39: New Approaches to Knowledge Acquisition (*Lu Ruqian*)

40: Current Trends in Theoretical Computer Science — Essays and Tutorials (Eds. *G Rozenberg & A Salomaa*)

41: Distributed Constraint Logic Programming (*Ho-Fung Leung*)

42: RLISP '88 — An Evolutionary Approach to Program Design & Reuse (*J Marti*)

43: Cooperation in Industrial Multi-agent Systems (*N Jennings*)

44: Compositional Methods for Communication Protocol Design — A Petri Net Approach (*N A Anisimov*)

45: Computer Simulation of Developing Structures in Nature, Society & Culture (*V V Alexandrov & A I Semenkov*)

46: Mathematical Aspects of Natural and Formal Languages (*G Paun*)

47: Mathematical Logic for Computer Science (2nd Edn.) (*Lu Zhongwan*)

For a complete list of published titles in the series, please write in to the publisher.

Series in Computer Science Vol. 47

MATHEMATICAL LOGIC FOR COMPUTER SCIENCE

Second Edition

Lu Zhongwan

Chinese Academy of Sciences
Beijing

Published by

World Scientific Publishing Co. Pte. Ltd.
P O Box 128, Farrer Road, Singapore 912805
USA office: Suite 1B, 1060 Main Street, River Edge, NJ 07661
UK office: 57 Shelton Street, Covent Garden, London WC2H 9HE

British Library Cataloguing-in-Publication Data
A catalogue record for this book is available from the British Library.

MATHEMATICAL LOGIC FOR COMPUTER SCIENCE (Second Edition)

ISBN 981-02-3091-5

This book is printed on acid-free paper.

Printed in Singapore by Uto-Print

PREFACE

Mathematical logic studies logical problems with mathematical methods, principally logical problems in mathematics. It is a branch of mathematics. There are two kinds of mathematical research, proof and computation, which are essentially related to each other. Hence mathematical logic is essentially related to computer science, and many branches of mathematical logic have applications in it. This book describes those aspects of mathematical logic which are closely related to each other, including classical and non-classical logics. Roughly, non-classical logics can be divided into two groups, those that rival classical logic and those which extend it. This first group includes, for instance, constructive logic and multi-valued logics. The second includes modal and temporal logics, etc. Of non-classical logics, this book chooses to describe constructive and modal logics.

Materials adopted in this book are intended to attend to both the peculiarities of logical systems and the requirements of computer science, but those concerning the applications of mathematical logic in computer science are not involved. Topics are discussed concisely with the essentials emphasized and the minor details excluded. For various logics, their background, language, semantics, formal deduction, soundness and completeness are the main topics introduced. Formal deduction is treated in the form of natural deduction which resembles ordinary mathematical reasoning.

This book consists of an introduction, nine chapters, and an appendix. In the Introduction, the nature of mathematical logic is explained. In Chapter 1 of prerequisites, the basic concepts of set theory, including the fundamental theorems of countable sets, are reviewed; and inductive definitions and proofs are explained in detail, since many concepts in mathematical logic are defined inductively. Besides these prerequisites, this book is self-contained.

Chapters 2–5 describe classical logics. Classical propositional logic may be regarded as part of classical first-order logic; but these logics are described separately in Chapters 2 and 3 because classical propositional logic has its own characteristics. Classical logic is the basis of this book; its soundness and completeness are studied in Chapter 5. Especially, the completeness problem of classical propositional logic and the different cases of classical first-order logic with and without equality are treated separately, in order to show the distinction of these cases in the treatment of completeness. Chapter 4 introduces the axiomatic deduction system, and proves the equivalence between it and the natural deduction system.

Chapter 6 studies, on the basis of soundness and completeness, Compactness Theorem, Löwenheim-Skolem Theorem, and Herbrand Theorem, which is the basis of one approach of automatic theorem proving in artificial intelligence.

Chapters 7–9 describe constructive and modal logics, and discuss the relationship between classical logic and these non-classical logics.

In Appendix, a simple form of formal proof in natural deduction system is introduced.

The first edition of this book was printed in 1989. The revisions in this edition are essentially concerned with rewriting proofs and expanding the explanations in the remarks. New terms and notations are adopted instead of original ones; for instance, "propositional logic" and "first-order logic" are renamed as "classical propositional logic" and "classical first-order logic", and "interpretation" and "assignment" are combined into one term "valuation". Furthermore, Sec. 6.4 of Chapter 6 is eliminated.

I would like to offer my deepest thanks to many people. Professor Hu Shihua taught me mathamatical logic selflessly. In the writing of this book, Professor Wang Shiqiang, Professor Tang Zhisong, Professor Xu Kongshi, Professor Yang Dongping, and the late Professor Wu Yunzeng provided much criticism and advice. Mr. Zhang Li helped in making suggestions and preparing the revisions. The Graduate School of University of Science and Technology of China (in Beijing) and Tsinghua University provided me with the opportunity to teach the materials of this book. The discussions with the students during my years of teaching in the universities have been very helpful in the revision of this book.

I would also like to thank the staff of World Scientific Publishing Company, first Professor K. K. Phua, and then Mr. S. J. Han, Ms. G. K. Tan, Ms. Jennifer Gan, Ms. H. M. Ho, and Ms. S. H. Gan, for their friendly and efficient help in the production of this book.

Finally I would like to express gratitude to my wife Ding Yi for her patient typing and encouragement during the long writing period.

Lu Zhongwan

Institute of Software, Chinese Academy of Sciences
Garduate School of University of Science and Technology of China (in Beijing)
October 1996

CONTENTS

Preface v

Introduction 1

1. Prerequisites 5

    1.1 Sets 5
    1.2 Inductive definitions and proofs 11
    1.3 Notations 15

2. Classical Propositional Logic 17

    2.1 Propositions and connectives 18
    2.2 Propositional language 21
    2.3 Structure of formulas 26
    2.4 Semantics 33
    2.5 Tautological consequence 40
    2.6 Formal deduction 45
    2.7 Disjunctive and conjunctive normal forms 61
    2.8 Adequate sets of connectives 65

3. Classical First-Order Logic 69

    3.1 Proposition functions and quantifiers 70
    3.2 First-order language 74
    3.3 Semantics 83
    3.4 Logical consequence 93
    3.5 Formal deduction 97
    3.6 Prenex normal form 106

4. Axiomatic Deduction System 109

4.1 Axiomatic deduction system 109
4.2 Relation between the two deduction systems 112

5. Soundness and Completeness 117

5.1 Satisfiability and validity 118
5.2 Soundness 125
5.3 Completeness of propositional logic 127
5.4 Completeness of first-order logic 133
5.5 Completeness of first-order logic with equality 137
5.6 Independence 140

6. Compactness, Löwenheim–Skolem and Herbrand Theorems 147

6.1 Compactness Theorem 147
6.2 Löwenheim–Skolem's Theorem 148
6.3 Herbrand's Theorem 149

7. Constructive Logic 157

7.1 Constructivity of proofs 157
7.2 Semantics 159
7.3 Formal deduction 163
7.4 Soundness 171
7.5 Completeness 172

8. Modal Propositional Logic 179

8.1 Modal propositional language 179
8.2 Semantics 180
8.3 Formal deduction 185
8.4 Soundness 192
8.5 Completeness of T 195
8.6 Completeness of S_4, B, S_5 198

9. Modal First-Order Logic 205

9.1 Modal first-order language 205

9.2 Semantics 206
9.3 Formal deduction 209
9.4 Soundness 211
9.5 Completeness 212
9.6 Equality 217

Appendix (a simple form of formal proof in natural deduction) 221

Bibliography 227

List of Symbols 229

Index 233

INTRODUCTION

Mathematical logic is the study of logical problems, principally the logical problems in mathematics.

The premises and conclusions in reasoning are propositions, which are either true or false. Some logicians prefer to speak of sentences (or statements) instead of propositions. Their motivations might be that a sentence is used as a unit of expression in natural languages and a proposition as what a sentence asserts.

Some conclusion is said to be deducible from some premises when the truth of the premises implies that of the conclusion. Such reasoning is called deductive. Hence, the study of deductive reasoning is the study of those kinds of premises and conclusions that are in the deducibility relation. The contents of this book belong to deductive logic, in which deductive reasoning is studied.

We first consider some examples. The premises and conclusion in

1) The sum of the digits of every multiple of 3 is a multiple of 3. (Premise)
The sum of the digits of 10^{10} is not a multiple of 3. (Premise)
10^{10} is not a multiple of 3. (Conclusion)

are true propositions and the reasoning in 1) is correct. The correctness seems to be concerned with the truth of the premises and conclusion. But this is not the case. The reasoning in

2) Every middle school student plays tennis. (Premise)
Z does not play tennis. (Premise)
Z is not a middle school student. (Conclusion)

is also correct, and the justification for its correctness is the same as that for the correctness of 1). But the premises and conclusion in 2) may be true or false. Besides, the matter of the propositions in 2) is quite distinct from that in 1).

Therefore the correctness of reasoning is neither concerned with the matter, nor with the truth or falsehood of the premises and conclusion. Deducibility requires only that the truth of the premises implies that of the conclusion. Mathematical logic does not study the truth or falsehood of the premises and conclusion, but studies whether the truth of the premises implies that of the conclusion.

Then, by what is the deducibility relation determined?

A proposition has its matter, which determines its truth or falsehood, and its logical form (or simply, form). It is the logical forms of the premises and conclusion which determine the deducibility relation between them.

The premises and conclusion in both 1) and 2) are of the following logical forms respectively:

3)
- Every member of S has the property P. (Premise)
- α does not have the property P. (Premise)
- α is not a member of S. (Conclusion)

Obviously, for any three propositions, if they are respectively of the logical forms in 3), then the last proposition can be deduced from the first two (no matter what set, property, and member S, P, and α are, respectively).

Mathematical logic is concerned with the analysis of the premises and conclusions with attention to the logical form in abstraction from the matter and from the truth or falsehood.

When propositions are expressed and their logical forms analysed in natural languages, confusion sometimes arises. For instance, in the following two arguments:

4)
- X knows Y. (Premise)
- Y is the captain of the football team. (Premise)
- X knows the captain of the football team. (Conclusion)

5)
- X knows somebody in Class A. (Premise)
- Somebody in Class A is the captain of the football team. (Premise)
- X knows the captain of the football team. (Conclusion)

the corresponding propositions are similar linguistically. But the argument in 4) is correct, while that in 5) is not. This illustrates that linguistic similarity in natural languages does not in general imply the sameness of logical form.

For these reasons we need to construct a kind of symbolic language to replace the natural languages. Such artificial symbolic language is called *formal language*, in which symbols are used to form formulas and formulas serve to express propositions. The logical forms of propositions can be expressed precisely by formulas.

As in the case of natural languages, formal language has its semantics and syntax. *Semantics* is concerned with the meaning of expressions when the symbols are interpreted in a certain way. *Syntax*, on the other hand, is concerned with the formal structure of expressions, irrespective of any interpretation. These two aspects of a formal language must be distinguished from each other.

Discussion of topics takes place in some language. But now, the object being discussed is itself a language. Hence two languages on different levels are involved. The language being discussed is called the *object language*, that is the formal language. The language in which the discussion takes place is called the *metalanguage*. The metalanguage used here is the English language.

Traditionally mathematics does not make the language of mathematics or its method of reasoning an object of study. Mathematical logic attempts to study these aspects mathematically (by first making precise the language and the inferences used). It thus becomes a new branch of mathematics.

It is customary to trace back to Leibniz (1646–1716) for the ideas of modern mathematical logic. Leibniz strove for an exact universal language of science and looked for a calculus of reasoning so that arguments and disagreements can be settled by calculation. These purposes were accomplished in Frege [1879], hence it is convenient to date the beginning of mathematical logic back to that year. Such exact language of science is the formal language to be constructed and such calculus of reasoning is the system of formal deducibility to be developed in the following chapters.

1
PREREQUISITES

The only prerequisite for reading this book is familiarity with the basic notions of sets, inductive definitions, and inductive proofs. Here a brief summary of these will be given. The reader may omit this chapter at first reading and refer to it when necessary.

1.1. SETS

A *set* is a collection of objects, called *members* or *elements*. We write

$$\alpha \in S$$

to mean that α is a member of S, and write

$$\alpha \notin S$$

to mean that α is not a member of S.

For convenience, we write

$$\alpha_1, \ldots, \alpha_n \in S$$

to mean that $\alpha_1 \in S, \ldots,$ and $\alpha_n \in S$, and write

$$\alpha_1, \ldots, \alpha_n \notin S$$

to mean that $\alpha_1 \notin S, \ldots,$ and $\alpha_n \notin S$.

Sets are determined by their members. Two sets S and T are said to be *equal*, written as

$$S = T$$

iff (the word "iff" is used as an abbreviation of "if and only if") they have the same members, that is, for every $x, x \in S$ iff $x \in T$.

$S \neq T$ means that S and T are unequal, that is, there is some x such that $x \in S$ iff $x \notin T$.

The totality of members contained in a set is called its extension. Hence a set is determined by its extension. The intension of a set is the common property of its members. For instance, the extension of the set of non-negative even numbers is

$$\{0, 2, 4, \dots\},$$

and its extension is "being a non-negative integer divisible by 2". The extension of the set $\{\alpha, \beta, \gamma\}$ is α, β, and γ, and its intension is "being α or β or γ".

S is said to be a *subset* of T, written as

$$S \subseteq T$$

iff for every $x, x \in S$ implies $x \in T$. Every set is a subset of itself. $S = T$ iff $S \subseteq T$ and $T \subseteq S$.

S is said to be a *proper subset* of T, iff $S \subseteq T$ and $S \neq T$.

A set with $\alpha_1, \dots, \alpha_n$ as its members is written as

$$\{\alpha_1, \dots, \alpha_n\} .$$

Obviously, we have

$$\begin{aligned}
&\{\alpha\} = \{\alpha, \alpha\} , \\
&\{\alpha, \beta\} = \{\beta, \alpha\} = \{\alpha, \beta, \beta\} = \{\alpha, \beta, \beta, \alpha\} , \\
&\{\alpha, \beta, \gamma\} = \{\alpha, \gamma, \beta\} = \{\gamma, \beta, \alpha\} = \{\alpha, \beta, \alpha, \gamma\} .
\end{aligned}$$

Hence the components of a set are independent of the order and repetition of its members.

One special set is the empty set $\emptyset$, which has no member at all. $\emptyset$ is a subset of any set S. $\emptyset \subseteq S$ is said to be vacuously true, since it requires doing nothing to verify that for any member $x \in \emptyset, x \in S$ also holds. Or in other words, $\emptyset \subseteq S$ is false iff there is some x such that $x \in \emptyset$ and $x \notin S$, which is impossible.

We write

$$\{x \mid \_\_x\_\_\}$$

for the set of all objects x such that $\_\_x\_\_$. For instance, suppose

$$S = \{x \mid x < 100 \text{ and } x \text{ is prime}\} \ ,$$
$$T = \{x \mid x = 0 \text{ or } x = 1 \text{ or } x = 2\} \ ,$$

then S is the set of all primes less than 100, and $T = \{0, 1, 2\}$.
The set $\{x \mid x \in S \text{ and } \_\_x\_\_\}$ may be written as

$$\{x \in S \mid \_\_x\_\_\} \ .$$

We define

$$\overline{S} = \{x \mid x \notin S\} \ ,$$
$$S \cup T = \{x \mid x \in S \text{ or } x \in T\} \ ,$$
$$S \cap T = \{x \mid x \in S \text{ and } x \in T\} \ ,$$
$$S - T = \{x \mid x \in S \text{ and } x \notin T\} \ .$$

$\overline{S}$ is called the *complement* of S; $S \cup T$, $S \cap T$ and $S - T$ are called the *union*, *intersection*, and *difference* of S and T respectively.

S and T are said to be *disjoint* iff $S \cap T = \emptyset$.

More generally, suppose $S_1, \dots, S_n$ are sets and $n \geq 2$, we set

$$S_1 \cup \ldots \cup S_n = \{x \mid x \in S_i \text{ for some } i = 1, \dots, n\} \ ,$$
$$S_1 \cap \ldots \cap S_n = \{x \mid x \in S_i \text{ for each } i = 1, \dots, n\} \ .$$

Suppose $\{S_i \mid i \in I\}$ is a collection of sets indexed by members of the set I. Then we set

$$\bigcup_{i \in I} S_i = \{x \mid x \in S_i \text{ for some } i \in I\} \ ,$$
$$\bigcap_{i \in I} S_i = \{x \mid x \in S_i \text{ for each } i \in I\} \ .$$

They are respectively the union and intersection of $\{S_i \mid i \in I\}$.

The *ordered pair* of objects α and β is written as

$$\langle \alpha, \beta \rangle \ .$$

Then $\langle \alpha, \beta \rangle = \langle \alpha_1, \beta_1 \rangle$ iff $\alpha = \alpha_1$ and $\beta = \beta_1$.

The *ordered n-tuple*

$$\langle \alpha_1, \ldots, \alpha_n \rangle$$

is the same as the finite sequence $\alpha_1, \ldots, \alpha_n$. Then $\langle \alpha_1, \ldots, \alpha_n \rangle = \langle \beta_1, \ldots, \beta_m \rangle$ iff $n = m$ and $\alpha_i = \beta_i$ for $i = 1, \ldots, n$.

A set of ordered n-tuples is also written with the notation

$$\{\langle x_1, \ldots, x_n \rangle | \_x_1\_, \ldots, \_x_n\_\} .$$

For instance,

$$\{\langle m, n \rangle | \, m, n \text{ are natural numbers and } m < n\}$$

is the set of ordered pairs of natural numbers of which the first component is smaller than the second.

The *Cartesian product* $S_1 \times \ldots \times S_n$ of sets $S_1, \ldots, S_n$ is defined by

$$S_1 \times \ldots \times S_n = \{\langle x_1, \ldots, x_n \rangle | \, x_1 \in S_1, \ldots, x_n \in S_n\} .$$

When $S_1, \ldots, S_n$ are the same, the nth Cartesian product S^n of S is

$$S^n = \underbrace{S \times \ldots \times S}_{n} = \{\langle x_1, \ldots, x_n \rangle | \, x_1, \ldots, x_n \in S\} .$$

For $n \geq 1$, an n-ary *relation* R on a set S is the set R:

$$R = \{\langle x_1, \ldots, x_n \rangle | x_1, \ldots, x_n \in S \text{ and relation } R \text{ exists among } x_1, \ldots, x_n \text{ in this order } \}.$$

Hence $R \subseteq S^n$. A unary relation R on S is a property:

$$R = \{x \in S \,|\, x \text{ has the property } R\} ,$$

which is a subset of S. A special binary relation on any set S is the equality relation:

$$\{\langle x, y \rangle \,|\, x, y \in S \text{ and } x = y\}$$

or

$$\{\langle x, x \rangle | \, x \in S\} .$$

It is a subset of S^2.

A relation (as a set) has its extension and intension. The intension of a relation is its meaning. Its extension is the set of all ordered n-tuples which are in this relation. For instance, the intension of the property (unary relation) "being even number" on the set of natural numbers is "divisible by 2", and its extension is $\{0, 2, 4, \dots\}$. The intension of the binary relation "$m < n$" on natural numbers is "there exists non-zero natural number x, such that $m + x = n$", and its extension is $\{\langle m, n\rangle | m$ and n are natural numbers, and $m < n\}$.

The extension and intension of a relation are different concepts. It is obvious that the relation concept defined above is its extension.

A *function* (*mapping*) f is a set of ordered pairs such that if $\langle x, y\rangle \in f$ and $\langle x, z\rangle \in f$, then $y = z$. The *domain* $dom(f)$ of f is the set

$$dom(f) = \{x \,|\, \langle x, y\rangle \in f \text{ for some } y\} ,$$

and the *range* $ran(f)$ of f is the set

$$ran(f) = \{y \,|\, \langle x, y\rangle \in f \text{ for some } x\} .$$

If f is a function and $x \in dom(f)$, then the unique y for which $\langle x, y\rangle \in f$ is denoted by $f(x)$ and is called the *value of f at x*. If f is a function with $dom(f) = S$ and $ran(f) \subseteq T$, we say that f is a *function from S to (into) T* (or *f maps S into T*), and denote it by

$$f : S \to T .$$

If in addition $ran(f) = T$, then *f maps S onto T* (f is a *surjection*). A function f is *one-one* (an *injection*) if $f(x) = f(y)$ implies $x = y$.

An n-ary function on S is a function mapping S^n into S. For instance, addition is a binary function on the set N of natural numbers, and the successor is a unary function on N. If f is n-ary and $\langle x_1, \dots, x_n\rangle \in dom(f)$, then we write

$$f(x_1, \dots, x_n)$$

for $f(\langle x_1, \dots, x_n\rangle)$.

Suppose R is an n-ary relation on S and $S_1 \subseteq S$. The *restriction of R to S_1* is the n-ary relation $R \cap S_1^n$.

Suppose $f : S \to T$ is a function and $S_1 \subseteq S$. The *restriction of f to S_1* is the function

$$f|S_1 : S_1 \to T$$

which is defined by

$$(f|S_1)(x) = f(x) \text{ for every } x \in S_1 .$$

Suppose R is a binary relation. We often write

$$xRy$$

for $\langle x, y \rangle \in R$.

We define the following notions.

R is *reflexive* on S, iff for any $x \in S$, xRx.

R is *symmetric* on S, iff for any $x, y \in S$, whenever xRy, then yRx.

R is *transitive* on S, iff for any $x, y, z \in S$, whenever both xRy and yRz, then xRz.

R is an *equivalence relation* on S, iff R is reflexive, symmetric, and transitive on S.

Suppose R is an equivalence relation on S. For any $x \in S$, the set

$$\overline{x} = \{y \in S | \, xRy\}$$

is called the *R-equivalence class of x*. Then the R-equivalence classes make a *partition* of S, that is, the R-equivalence classes are subsets of S such that each member of S belongs to exactly one R-equivalence class. Then for any $x, y \in S$,

$$\overline{x} = \overline{y} \text{ iff } xRy .$$

The cardinal of a set S is a measure of its size. For finite sets, natural numbers can be used as measures of size. Cardinals generalize this situation to infinite sets.

Two sets S and T are said to be *equipotent*, written as

$$S \sim T$$

iff there is a one-one function from S onto T. Obviously $\sim$ is an equivalence relation. This permits the classification of sets with respect to the notion of equipotence, and thus we can generalize the notion of the number of members in a set so that it covers infinite sets.

A *cardinal* (or *cardinal number, power*) of a set S, denoted by

$$|S| ,$$

is associated with S in such a way that

$$|S| = |T| \quad \text{iff} \quad S \sim T\ .$$

Then two sets have the same cardinal iff they are equipotent. When S is finite, some natural number is taken to be $|S|$. A finite set S is equipotent to $\{0, \ldots, n-1\}$ for some natural number n. We note that $|\emptyset| = 0$

By $|S| \leq |T|$ we mean that there is a one-one function from S into T.

S is said to be *countably* (or *enumerably*) *infinite*, iff $|S| = |N|$. S is said to be *countable* (or *enumerable*), iff $|S| \leq |N|$, that is, iff S is finite or countably infinite.

We state several theorems about countable sets with their proofs omitted.

Theorem 1.1.1.
A subset of a countable set is countable. □

Theorem 1.1.2.
The union of any finite number of countable sets is countable. □

Theorem 1.1.3.
The union of countably many countable sets is countable. □

Theorem 1.1.4.
The Cartesian product of any finite number of countable sets is countable. □

Theorem 1.1.5.
The set of all finite sequences with the members of a countable set as components is countable. □

1.2. INDUCTIVE DEFINITIONS AND PROOFS

Inductive definitions of sets are often presented informally by giving some rules for generating members of the set and then adding that an object is to be in the set only if it has been generated according to the rules. An equivalent formulation of the definition is to characterise the set as the smallest one closed under the rules.

The basic example of an inductive definition is that of the set N of natural numbers.

Definition 1.2.1.
[1] $0 \in N$.
[2] For any n, if $n \in N$, then $n' \in N$ (n' being the successor of n).
[3] $n \in N$ only if n has been generated by [1] and [2].

Definition 1.2.1 can be formulated equivalently as follows.

Definition 1.2.2.
N is the smallest set S such that
[1] $0 \in S$.
[2] For any n, if $n \in S$, then $n' \in S$.

Suppose R is a property and $R(x)$ means that x has property R.

Theorem 1.2.3.
If
[1] $R(0)$.
[2] For any $n \in N$, if $R(n)$, then $R(n')$.
then $R(n)$ for any $n \in N$.

Proof. Suppose $S = \{n \in N \mid R(n)\}$. Then S satisfies [1] and [2] of Definition 1.2.2. Hence $N \subseteq S$, that is, $R(n)$ for any $n \in N$. □

A proof by means of Theorem 1.2.3 is called an *inductive proof* or a *proof by induction*. In connection with proofs by induction, we shall use the following terminology. The proposition "For every $n \in N$, $R(n)$" is the *induction proposition*, and the variable n in it is the *induction variable*. The proof consists of two steps. The first step, called the *basis* of induction, is the proof of [1] of Theorem 1.2.3. The second step, called the *induction step*, is the proof of [2]. The assumption $R(n)$ in the induction step is called the *induction hypothesis*. For the sake of simplicity we shall sometimes write "ind hyp" for the induction hypothesis.

The condition [2] in Theorem 1.2.3 may be replaced by

[2°] For any $n \in N$, if $R(0), \ldots, R(n)$, then $R(n')$.

That is, "For any $n \in N$, $R(n)$" can be derived from [1] and [2°] as well. This is another version of proof by induction, called *course-of-values induction.*

The proof of course-of-values induction is as follows. Let $S = \{n \in N | R(0) \text{ and } \ldots \text{ and } R(n)\}$. By [1] and $0 \in N$ (see Definition 1.2.1 [1]), we have $0 \in S$. Suppose $n \in S$, that is, $n \in N$ and $R(0), \ldots, R(n)$. By [2], we obtain $R(n')$. By Definition 1.2.1 [2], $n' \in N$ follows from $n \in N$. Then $n' \in S$. Thus S satisfies [1] and [2] of Definition 1.2.2, and $N \subseteq S$. Hence for any $n \in N, R(n)$.

Course-of-values induction has still another version, in which the following [2*] is used instead of [1] and [2°]:

[2*] For any $n \in N$, if $R(m)$ for each $m < n$, then $R(n)$.

When $n = 0$, [2*] is

1) If $R(m)$ for each $m < 0$, then $R(0)$.

Because "$m < 0$" is false, "$R(m)$ for each $m < 0$" (which means "if $m < 0$, then $R(m)$") is vacuously true. Then, by 1), we have $R(0)$, which is [1]. [2°] follows obviously from [2*]. Therefore "For every $n \in N, R(n)$" can be derived from [2*].

Recursion is a method of defining a function on an inductively defined set by specifying each of its values in terms of previously defined values, employing already given functions.

For instance, let g and h be given functions on N, then the following equations:

$$\begin{cases} f(0) = g(0) \\ f(n') = h(f(n)) \end{cases}$$

define a function f on N from g and h. Although at first sight f seems to be defined in terms of itself, yet this is not the case, because, for every $n \in N, f(n)$ can be computed in finite steps from the defining equations. Such a definition is called a *definition by recursion.*

Theorem 1.2.4. (***Principle of definition by recursion***)

Suppose g and h are given functions on N. Then there exists a unique function f satisfying

$$\begin{cases} f(0) = g(0) \ , \\ f(n') = h(f(n)) \ . \end{cases}$$

Proof. By induction on n. □

The second equation in the definition by recursion may be of the following form:

$$f(n') = h(n, f(n))$$

where h is a binary function of n and $f(n)$, n being the number of times of applying the second equation before $f(n')$ is computed.

The general case of an inductive definition of a set S is as follows. Suppose a set M and n_i-ary functions g_i $(i = 1, \ldots, k)$ are given. S is the smallest set T such that $M \subseteq T$ and, for any $x_1, \ldots, x_{n_i} \in T$, $g_i(x_1, \ldots, x_{n_i}) \in T$.

Then, in an inductive proof of a proposition that every member of S has a certain property, the basis of induction is to prove that every member of S which is generated outright (that is, members of the given set M) has this property. The induction step is to prove that the given functions g_i preserve this property, that is, when certain members of S have this property, the members generated from them by means of g_i have this property as well.

The general case of the principle of definition by recursion is as follows. Suppose S is the inductively defined set described above. Let

$$\begin{aligned} &h : M \to S \\ &h_i : S^{n_i} \to S \quad (i = 1, \ldots, k) \end{aligned}$$

be given functions. Then there exists a unique function f such that

$$\begin{cases} f(x) = h(x) & \text{for any } x \in M \ , \\ f(g_i(x_1, \ldots, x_{n_i})) = h_i(f(x_1), \ldots, f(x_{n_i})) & \text{for any } x_1, \ldots, x_{n_i} \in S \ . \end{cases}$$

We note that the generation of any member in the set N of natural numbers is unique, but generally, the members in an inductively defined set S are not necessarily uniquely generated. For instance, given $M = \{0, 1\}$ and unary function g such that $g(0) = 1$ and $g(1) = 0$. Then $S = \{0, 1\}$. The member $0 \in S$ is not uniquely generated: $0 \in M$ or $0 = g(1)$. When the definition by recursion of f and the principle of definition by recursion are involved, the uniqueness of generation of the members in S is required.

Inductive definitions and definitions by recursion will be used repeatedly in the description of the syntax and semantics of formal languages.

1.3. NOTATIONS

The following standard conventions in mathematics will be used throughout this book.

The symbol $\Longrightarrow$ is used for "implies", and $\Longleftrightarrow$ for "iff". We also use $\Longleftarrow$ for the converse of $\Longrightarrow$.

Suppose $\mathcal{A}_1, \ldots, \mathcal{A}_n$ are propositions. We write

$$\mathcal{A}_1 \Longrightarrow \mathcal{A}_2 \Longrightarrow \ldots \Longrightarrow \mathcal{A}_n$$

for "$\mathcal{A}_1 \Longrightarrow \mathcal{A}_2, \ldots, \mathcal{A}_{n-1} \Longrightarrow \mathcal{A}_n$", and

$$\mathcal{A}_1 \Longleftrightarrow \mathcal{A}_2 \Longleftrightarrow \ldots \Longleftrightarrow \mathcal{A}_n$$

for "$\mathcal{A}_1 \Longleftrightarrow \mathcal{A}_2, \ldots, \mathcal{A}_{n-1} \Longleftrightarrow \mathcal{A}_n$".

"Def", "Thm", "Lem", and "Cor" are abbreviations of "Definition", "Theorem", "Lemma", and "Corollary" respectively.

Each chapter is divided into sections. Definitions and theorems (including lemmas and corollaries) in each section are numbered consecutively. For instance, a reference such as "Definition 2.2.3" means the third numbered item which is a definition, in Section 2.2 of Chapter 2. Exercises in each section have another system of numbering.

For reference, certain formulas and statements in a section are denoted by "1)", "2)", etc., and those in a proof or in an example are denoted by "(1)", "(2)", etc.

The symbol □ is used to denote the end of a proof or, when it appears immediately after a theorem, to indicate that the proof is immediate and accordindly omitted.

Reference to the bibliography is made by citing the author and the year of publication of the work.

2
CLASSICAL PROPOSITIONAL LOGIC

Classical logic is to be introduced in Chapters 2–5. According to the viewpoint of classical logic, a proposition is either true or false. Truth and falsehood are values of a proposition. A proposition takes one of truth and falsehood as its value. For any proposition $\mathcal{A}$, the proposition "$\mathcal{A}$ or not $\mathcal{A}$" is true. Classical propositional logic is first introduced in this chapter.

Propositional logic is a part of mathematical logic. It includes only a part of logical forms and principles.

In propositional logic, compound propositions are composed from simple ones (as basic units) by using connectives. The characteristic of propositional logic is that, in studying the logical forms of propositions, only the logical forms of compound propositions are analysed to see how they are composed from initial components — simple propositions, while the logical forms of simple propositions are not analysed. In propositional logic, simple propositions are taken as a whole, which are either true or false.

For instance, the following

From "$\mathcal{A}$ or $\mathcal{B}$" and "not $\mathcal{A}$", $\mathcal{B}$ is deduced.

is a correct inference, where "$\mathcal{A}$ or $\mathcal{B}$" and "not $\mathcal{A}$" are logical forms of compound propositions. To see the correctness of the above inference, we need not analyse $\mathcal{A}$ and $\mathcal{B}$, because the correctness is determined by the logical forms of compound propositions.

Propositional logic studies the deducibility relations between premises and conclusions which are compound propositions or unanalysed simple propositions. The logical forms of compound propositions are determined by connectives. Hence propositional logic may also be called the logic of connectives.

2.1. PROPOSITIONS AND CONNECTIVES

Propositions formed by means of connectives are called *compound* propositions. The connectives most commonly used are "*not*", "*and*", "*or*", "*if, then*", and "*iff*". "Not" is unary, while the other four are binary.

The following are some examples of compound propositions:

1) 2 is not odd. (Not that 2 is odd.)
2) 2 is even and prime. (2 is even and 2 is prime.)
3) If a pair of opposite sides of a quadrilateral are parallel and equal, then it is a parallelogram.

The components of a compound proposition may or may not still be compound. For example, the component of 1) is "2 is odd", which is not compound, while one component of 3), "a pair of opposite sides of a quadrilateral are parallel and equal", is still a compound proposition.

The initial components of compound propositions are not compound. Non-compound propositions are called *simple* propositions. Simple propositions are not formed by means of connectives.

A proposition is either true or false. Truth or falsehood is the *value* (or *truth value*) of a proposition. The value of a true proposition is *truth*, that of a false one is *falsehood*. Usually truth is denoted by "1" and falsehood by "0".

The value of a compound proposition is determined by the values of its components and the connectives used. Let $\mathcal{A}$ and $\mathcal{B}$ be arbitrary propositions. The following compound propositions are formed by the common connectives:

Not $\mathcal{A}$.

$\mathcal{A}$ and $\mathcal{B}$.

$\mathcal{A}$ or $\mathcal{B}$.

If $\mathcal{A}$ then $\mathcal{B}$.

$\mathcal{A}$ iff $\mathcal{B}$.

We shall consider how the values of these compound propositions are determined.

Obviously $\mathcal{A}$ is true iff "not $\mathcal{A}$" is false. The meaning of $\mathcal{A}$ is irrelevant. The situation can be described by the following table:

$\mathcal{A}$	not $\mathcal{A}$
1	0
0	1

"$\mathcal{A}$ and $\mathcal{B}$" is true iff both $\mathcal{A}$ and $\mathcal{B}$ are true. Hence we have the following table:

$\mathcal{A}$	$\mathcal{B}$	$\mathcal{A}$ and $\mathcal{B}$
1	1	1
1	0	0
0	1	0
0	0	0

We have in the table one row for each of the possible combinations of values of $\mathcal{A}$ and $\mathcal{B}$. The last column gives the corresponding values of "$\mathcal{A}$ and $\mathcal{B}$".

According to the usual meaning of the word "or", "$\mathcal{A}$ or $\mathcal{B}$" is true when one of $\mathcal{A}$ and $\mathcal{B}$ is true, and is false when both $\mathcal{A}$ and $\mathcal{B}$ are false. When both $\mathcal{A}$ and $\mathcal{B}$ are true, the value of "$\mathcal{A}$ or $\mathcal{B}$" is to be determined according to the interpretation of "$\mathcal{A}$ or $\mathcal{B}$" adopted. It may be interpreted in the inclusive sense of "$\mathcal{A}$ or $\mathcal{B}$ or both", or in the exclusive sense of "$\mathcal{A}$ or $\mathcal{B}$ but not both". In mathematics the inclusive sense of "or" is commonly used, hence the values of "$\mathcal{A}$ or $\mathcal{B}$" are determined as follows:

$\mathcal{A}$	$\mathcal{B}$	$\mathcal{A}$ or $\mathcal{B}$
1	1	1
1	0	1
0	1	1
0	0	0

"If $\mathcal{A}$ then $\mathcal{B}$" (or "$\mathcal{A}$ implies $\mathcal{B}$") calls for more explanations.

The English words "if, then" and "imply" (or their translations in other natural languages), as used in everyday speech, seem often to denote a relation between the propositions they connect. Their possible meanings when employed in this way are difficult to fix precisely. One use of these words is adopted here, in which "if $\mathcal{A}$ then $\mathcal{B}$" means "if $\mathcal{A}$ is true then $\mathcal{B}$ is true" or "not that $\mathcal{A}$ is true and $\mathcal{B}$ is false". According to this meaning, the values of "if $\mathcal{A}$ then $\mathcal{B}$" is determined by the table:

$\mathcal{A}$	$\mathcal{B}$	if $\mathcal{A}$ then $\mathcal{B}$
1	1	1
1	0	0
0	1	1
0	0	1

The first and second rows in the table are obvious. In the other two rows, since $\mathcal{A}$ is false, "not that $\mathcal{A}$ is true and $\mathcal{B}$ is false" is true. Hence "if $\mathcal{A}$ then $\mathcal{B}$" is true when $\mathcal{A}$ is false.

The difficulty arises with the value 1 assigned to "if $\mathcal{A}$ then $\mathcal{B}$" in the cases where $\mathcal{A}$ is false. Consideration of examples of implicational propositions "if $\mathcal{A}$ then $\mathcal{B}$" in which $\mathcal{A}$ is false might perhaps lead one to the conclusion that such propositions do not have a value at all. One might also gain the impression that such propositions are not useful or meaningful.

However, we shall be interested in deduction and proof, principally in mathematics. In this context the significance of an implicational proposition "if $\mathcal{A}$ then $\mathcal{B}$" is that its truth enables the truth of $\mathcal{B}$ to be inferred from the truth of $\mathcal{A}$, and nothing in particular to be inferred from the falsehood of $\mathcal{A}$. A very common sort of mathematical proposition can serve to illustrate this. For instance, the following proposition

4) $$\text{If } x > 3, \text{ then } x^2 > 9.$$

is true, irrespective of the value taken by x. Different values of x give rise to all possible combinations of truth values for "$x > 3$" and "$x^2 > 9$" except that combination "truth and falsehood". Taking $x = 4, -4, -3$ respectively yields the combinations "truth and truth", "falsehood and truth", "falsehood and falsehood", and these are the circumstances which, according to the above table of the truth values of "if $\mathcal{A}$ then $\mathcal{B}$", give 4) the truth value 1. The combination "truth and falsehood" is impossible, because 4) is true. The point to remember is that the only circumstance in which "if $\mathcal{A}$ then $\mathcal{B}$" is false is when $\mathcal{A}$ is true and $\mathcal{B}$ is false.

In set theory we have verified that $\emptyset \subseteq S$ is vacuously true for any set S. $\emptyset \subseteq S$ means:

$$\text{For all } x, \text{ if } x \in \emptyset, \text{ then } x \in S.$$

This is true since "$x \in \emptyset$" is false.

Generally, whenever $\mathcal{A}$ is false, "if $\mathcal{A}$ then $\mathcal{B}$" is vacuously true, since in such case the verification of "if $\mathcal{A}$ then $\mathcal{B}$" does not require doing anything to deduce $\mathcal{B}$ from $\mathcal{A}$.

Such use of "if, then" as illustrated above is familiar in mathematics. Although it may seem unusual, it yields no inconsistency with everyday speech. For instance, somebody may say: "If Z comes, then the sun rises in the West." Of course, the speaker understands that "Z comes" has no connection with "the sun rises in the West". What he intends to assert is that "Z comes" is false. Since he is sure of the falsehood of "Z comes", his whole proposition is true.

"$\mathcal{A}$ iff $\mathcal{B}$" is the same as "if $\mathcal{A}$ then $\mathcal{B}$, and if $\mathcal{B}$ then $\mathcal{A}$". Hence its truth values are determined by the table:

$\mathcal{A}$	$\mathcal{B}$	$\mathcal{A}$ iff $\mathcal{B}$
1	1	1
1	0	0
0	1	0
0	0	1

An (n-ary) function with the set of all ordered (n-) tuples of truth values as its domain and the set $\{1, 0\}$ as its range is called an (n-ary) *truth function*. Thus, the connectives are truth functions. "Not" is unary; "and", "or", "if, then", and "iff" are binary truth functions.

2.2. PROPOSITIONAL LANGUAGE

In this section the *propositional language* $\mathcal{L}^p$ is to be constructed. It is the formal language for propositional logic.

A formal language is a collection of symbols, which should be distinguished from symbols of the metalanguage used in studying them.

$\mathcal{L}^p$ consists of three classes of symbols. The first class includes an infinite sequence of *proposition symbols*. We use the roman-type small Latin letters:

$$\mathrm{p} \quad \mathrm{q} \quad \mathrm{r}$$

(with or without subscripts or superscripts) to denote arbitrary proposition symbols. The second kind includes five *connective symbols*, or simply *connectives*:

$$\neg \quad \wedge \quad \vee \quad \rightarrow \quad \leftrightarrow$$

Their oral reading and English names in standard use are respectively as follows:

Oral reading	English name
not	*negation*
and	*conjunction*
or (and/or)	(*inclusive*) *disjunction*
if, then (imply)	*implication*
iff (be equivalent to)	*equivalence*

The third class includes two *punctuation symbols*, or simply *punctuation*:

$$(\quad)$$

which are called *left* and *right parentheses*.

The infinite sequence of proposition symbols are not specified. p, q, r, etc. are arbitrary members in the sequence. For instance, p may be the first, or the fifteenth, or the thirty-seventh,etc. Similarly for q and r. Hence p and q may be different or the same. But different occurrences of p in the same context must be the same proposition symbol.

Expressions are finite strings of symbols. For instance, p, pq, (r), $\mathrm{p}\wedge\rightarrow$ q, and $\neg(\mathrm{p}\vee\mathrm{q})$ are expressions of $\mathcal{L}^p$.

The *length* of an expression is the number of occurrences of symbols in it. The lengths of the five expressions given above are 1, 2, 3, 4, and 6, respectively.

There is one special expression of length 0. It is the *empty expression*, which cannot be written. The empty expression is analogous to the empty set. Therefore the notation, $\emptyset$, for the empty set is used for the empty expression.

Two expressions U and V are *equal*, written as $\mathrm{U} = \mathrm{V}$, iff they are of the same length and have the same symbols in order.

Unless otherwise stated, the scanning of symbols in expressions proceeds from left to right.

The expression formed by concatenating two expressions U and V in this order is denoted by UV. Similarly for three or more expressions. Obviously, $U\emptyset = \emptyset U = U$ for any expression U.

If $U = W_1VW_2$, where U, V, W_1, and W_2 are expressions, then V is a *segment* of U. If V is a segment of U and $V \neq U$, then V is a *proper segment* of U. Every expression is a segment of itself. The empty expression is a segment of every expression.

If $U = VW$, where U, V, and W are expressions, then V is an *initial segment* of U. If W is non-empty, then V is a *proper initial segment* of U. Similarly, W is a *terminal segment* of U, and it is a *proper* one if V is non-empty.

Atoms (or atomic formulas) and formulas are to be defined from expressions. Formulas (also called well-formed formulas) correspond to grammatically correct sentences in natural languages.

The sets of atoms and formulas of $\mathcal{L}^p$ are denoted by $Atom(\mathcal{L}^p)$ and $Form(\mathcal{L}^p)$ respectively.

Definition 2.2.1. ($Atom(\mathcal{L}^p)$)

$Atom(\mathcal{L}^p)$ is the set of expressions of $\mathcal{L}^p$ consisting of a proposition symbol only.

In this and the next section, the symbol $*$ is used for any one of the four binary connectives.

Definition 2.2.2. ($Form(\mathcal{L}^p)$)

An expression of $\mathcal{L}^p$ is a member of $Form$ ($\mathcal{L}^p$) iff its being so follows from [1]–[3]:

[1] $Atom(\mathcal{L}^p) \subseteq Form(\mathcal{L}^p)$.

[2] If $A \in Form(\mathcal{L}^p)$, then $(\neg A) \in Form(\mathcal{L}^p)$.

[3] If $A, B \in Form(\mathcal{L}^p)$, then $(A * B) \in Form(\mathcal{L}^p)$.

[1]–[3] in Definition 2.2.2 are the *formation rules* of formulas of $\mathcal{L}^p$. We may also say that an expression of $\mathcal{L}^p$ is a formula of $\mathcal{L}^p$ iff it can be generated by (a finite number of applications of) the formation rules.

The above definition can be formulated equivalently as follows.

Definition 2.2.3. (*Form*($\mathcal{L}^p$))

Form($\mathcal{L}^p$) is the smallest class of expressions of $\mathcal{L}^p$ closed under the formation rules of formulas of $\mathcal{L}^p$.

Example

The expression

$$((\mathrm{p} \vee \mathrm{q}) \to ((\neg \mathrm{p}) \leftrightarrow (\mathrm{q} \wedge \mathrm{r})))$$

is a formula, which can be generated as follows:

(1) p (by Def 2.2.2 [1]).
(2) q (by Def 2.2.2 [1]).
(3) $(\mathrm{p} \vee \mathrm{q})$ (by Def 2.2.2 [3], (1), (2)).
(4) $(\neg \mathrm{p})$ (by Def 2.2.2 [2], (1)).
(5) r (by Def 2.2.2 [1]).
(6) $(\mathrm{q} \wedge \mathrm{r})$ (by Def 2.2.2 [3], (2), (5)).
(7) $((\neg \mathrm{p}) \leftrightarrow (\mathrm{q} \wedge \mathrm{r}))$ (by Def 2.2.2 [3], (4), (6)).
(8) $((\mathrm{p} \vee \mathrm{q}) \to ((\neg \mathrm{p}) \leftrightarrow (\mathrm{q} \wedge \mathrm{r})))$ (by Def 2.2.2 [3], (3), (7)).

The generation of this formula from p, q and r by applications of the formation rules can be illustrated more clearly by the following tree:

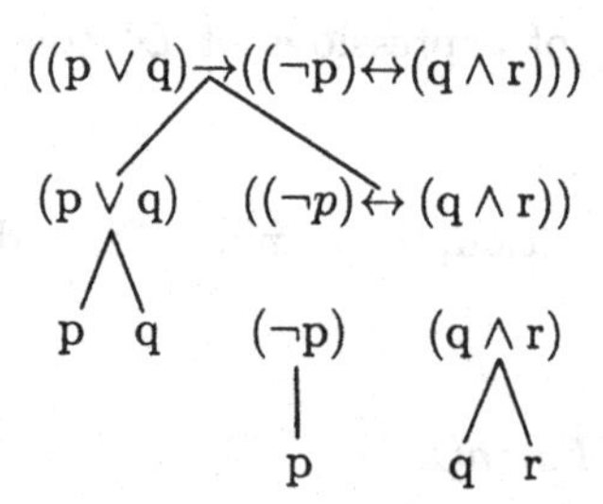

It is obvious that, in generating a formula, we obtain at each step a formula, which is a segment of the formula generated.

In the generation of the above formula, the order of Steps (3) and (4) may be exchanged, because $(\mathrm{p} \vee \mathrm{q})$ is not a segment of $(\neg \mathrm{p})$ and nor is $(\neg \mathrm{p})$ a segment of $(\mathrm{p} \vee \mathrm{q})$. Similarly (5) may be placed before (3) and (4). But (1) and (2) must be placed before (3), and (5) must be placed before (6).

We use the roman-type capital Latin letters:

$$\mathrm{A} \quad \mathrm{B} \quad \mathrm{C}$$

(with or without subscripts or superscripts) for any formula.

A, B, C, etc. may be different formulas or the same. Different occurrences of A in the same context must be the same formula. Such explanations will be omitted later.

Definitions 2.2.2 and 2.2.3 are inductive definitions, hence we have the following Theorem 2.2.4. (Compare them with Definitions 1.2.1 and 1.2.2 and Theorem 1.2.3 in the previous chapter.)

Theorem 2.2.4.

Suppose R is a property. If

[1] For any $\mathrm{p} \in Atom(\mathcal{L}^p)$, $R(\mathrm{p})$.

[2] For any $\mathrm{A} \in Form(\mathcal{L}^p)$, if $R(\mathrm{A})$, then $R((\neg \mathrm{A}))$.

[3] For any $\mathrm{A}, \mathrm{B} \in Form(\mathcal{L}^p)$, if $R(\mathrm{A})$ and $R(\mathrm{B})$, then $R((\mathrm{A} * \mathrm{B}))$.

then $R(\mathrm{A})$ for any $\mathrm{A} \in Form(\mathcal{L}^p)$. □

Applying Theorem 2.2.4, we can prove all formulas of $\mathcal{L}^p$ have certain property R. This is an inductive proof. The basis of induction is to prove any atomic formula has property R. The induction step is to prove the formulas generated by using connectives preserve property R. That is, for any formula A, suppose it has property R (induction hypothesis), we prove $(\neg A)$ has property R; for any formulas A and B, suppose they have property R (induction hypothesis), we prove $(A * B)$ has property R.

The above inductive proof is called *a proof by induction on the structure of the generation of formulas of* $\mathcal{L}^p$, or simply, *a proof by induction on the structure of formulas of* $\mathcal{L}^p$.

$Form(\mathcal{L}^p)$ is countable, since the set of symbols of $\mathcal{L}^p$ is countable and formulas are finite in length. (Refer to Section 1.1.)

Formulas have some important structural properties, which will be discussed in the next section.

Exercises 2.2.

2.2.1. $\mathrm{A}_1, \ldots, \mathrm{A}_n$ is a *formation sequence* of A iff $\mathrm{A}_n = \mathrm{A}$ and for $k \leq n$, A_k satisfies one of the following:

[1] $\mathrm{A}_k \in Atom(\mathcal{L}^p)$.

[2] $\mathrm{A}_k = (\neg \mathrm{A}_i)$ for some $i < k$.

[3] $\mathrm{A}_k = (\mathrm{A}_i * \mathrm{A}_j)$ for some $i, j < k$.

Show that an expression is a formula of $\mathcal{L}^p$ iff it has a formation sequence.

2.2.2. Suppose the number of occurrences of atoms in a formula A is m and that of occurrences of $\wedge, \vee, \rightarrow$, and $\leftrightarrow$ is n. Show that $m = n + 1$.

2.2.3. The *degree of complexity* of A $\in Form(\mathcal{L}^p)$ is defined by recursion:

$$\begin{cases} deg(\mathrm{A}) = 0 \;\text{ for atom A.} \\ deg((\neg \mathrm{A})) = deg(\mathrm{A}) + 1. \\ deg((\mathrm{A} * \mathrm{B})) = max(deg(\mathrm{A}), deg(\mathrm{B})) + 1. \end{cases}$$

[1] Show that $deg(\mathrm{A}) \leq$ the number of occurrences of connectives in A.

[2] Give examples of A such that $<$ or $=$ holds in [1].

2.2.4. Translate the following propositions into formulas (use atoms for simple propositions):

[1] He is clever and diligent.
[2] He is clever but not diligent.
[3] He didn't write the letter, or the letter was lost.
[4] He must study hard, otherwise he will fail.
[5] He will fail, unless he studies hard.
[6] He will go home, unless it rains.
[7] He will go home, only if it rains.
[8] If it rains, he will be at home; otherwise he will go to the market or school.
[9] The sum of two numbers is even iff both numbers are even or both numbers are odd.
[10] If y is an integer then z is not real, provided that x is rational.

2.3. STRUCTURE OF FORMULAS

In this section some structural properties of formulas will be discussed. The reader may omit the proofs at first reading, and understand the properties intuitively from the examples.

Lemma 2.3.1.

Every formula of $\mathcal{L}^p$ has the same number of occurrences of left and right parentheses. □

Lemma 2.3.2.

Any non-empty proper initial segment of a formula of $\mathcal{L}^p$ has more occurrences of left than right parentheses. Any non-empty proper terminal segment of a formula of $\mathcal{L}^p$ has less occurrences of left than right parentheses. Thus neither a non-empty proper initial segment nor a non-empty proper terminal segment of a formula can itself be a formula of $\mathcal{L}^p$.

Proof. By induction on the structure of formulas. □

Theorem 2.3.3.

Every formula of $\mathcal{L}^p$ is of exactly one of the six forms: an atom, $(\neg A)$, $(A \wedge B)$, $(A \vee B)$, $(A \to B)$, or $(A \leftrightarrow B)$; and in each case it is of that form in exactly one way.

Proof. This theorem consists of four parts:

(1) Every formula is of one of the six forms.
(2) Any two of the six forms are not the same.
(3) If $(\neg A) = (\neg A_1)$, then $A = A_1$.
(4) If $(A * B) = (A_1 * B_1)$, then $A = A_1$ and $B = B_1$.

(1) is obvious from Definition 2.2.2.

Proof of (2). An atom is a single symbol, hence it is different from the other five forms. Suppose

$$(\neg A) = (B * C).$$

Delete the first symbol on each side, obtaining

$$\neg A) = B * C).$$

Then B begins with $\neg$, which is impossible. Hence $(\neg A)$ is different from $(B * C)$. Suppose

$$(A \wedge B) = (A_1 \vee B_1).$$

We obtain

$$A \wedge B) = A_1 \vee B_1).$$

Then A and A_1 begin with the same occurrence of a symbol, and we must have $A = A_1$, otherwise one of A and A_1 will be a proper initial segment of the other, contradicting Lemma 2.3.2. Thus $\wedge$ and $\vee$ are identical, which is impossible. Hence $(A \wedge B)$ is different from $(A_1 \vee B_1)$. Similarly for any two binary connectives.

Proof of (3). If $(\neg A) = (\neg A_1)$, then obviously $A = A_1$.

Proof of (4). If $(A * B) = (A_1 * B_1)$, then $A = A_1$ as in the proof of (2), and accordingly the $*$'s on both sides are the same occurrence. Hence $B = B_1$. □

Example

Suppose $C = ((p \vee q) \to ((\neg p) \leftrightarrow (q \wedge r)))$. C is of the form of $(A \to B)$:

$$(\underbrace{(p \vee q)}_{A} \to \underbrace{((\neg p) \leftrightarrow (q \wedge r))}_{B}).$$

That is, C is generated from A and B by the $\to$ between them. Suppose C is generated by $\vee$, $\leftrightarrow$, or $\wedge$, that is,

$$C = (U \vee V) = (\underbrace{(p}_{U} \vee \underbrace{q) \to ((\neg p) \leftrightarrow (q \wedge r))}_{V}),$$

$$C = (U_1 \leftrightarrow V_1) = (\underbrace{(p \vee q) \to ((\neg p)}_{U_1} \leftrightarrow \underbrace{(q \wedge r))}_{V_1}),$$

$$C = (U_2 \wedge V_2) = (\underbrace{(p \vee q) \to ((\neg p) \leftrightarrow (q}_{U_2} \wedge \underbrace{r))}_{V_2}).$$

Then U, V, U_1, V_1, U_2, and V_2 are not formulas, since the numbers of occurrences of left and right parentheses in these expressions are not the same (by Lemma 2.3.1). Besides, C is not an atom, nor can it be generated by the $\neg$ in it. Hence, C can be of the form of $(A \to B)$ only.

Consider the formula $((p \to q) \to (p \to r))$. By the above arguments, it can be generated from $(p \to q)$ and $(p \to r)$ by the $\to$ between them, but not by the other $\to$'s. Hence it is of that form in exactly one way.

Remarks.

By Theorem 2.3.3, the generation of formulas of $\mathcal{L}^p$ is unique, if the order of certain steps in it is not considered. (See the explanations in the example after Definition 2.2.3.)

Since Theorem 2.3.3 asserts the uniqueness of each of these forms, we have the following definition.

Definition 2.3.4. (***Negation, Conjunction, Disjunction, Implication, Equivalence***)

$(\neg A)$ is called a *negation (formula)*. It is the negation of A.

$(A \wedge B)$ is called a *conjunction (formula)*. It is the conjunction of A and B. A and B are called the *conjuncts* of $(A \wedge B)$.

$(A \vee B)$ is called a *disjunction (formula)*. It is the disjunction of A and B. A and B are called the *disjuncts* of $(A \vee B)$.

$(A \rightarrow B)$ is called an *implication (formula)*. It is the implication of A and B. A and B are called the *antecedent* and *consequent* of $(A \rightarrow B)$.

$(A \leftrightarrow B)$ is called an *equivalence (formula)*. It is the equivalence of A and B.

Definition 2.3.5. (***Scope***)

If $(\neg A)$ is a segment of C, then A is called the *scope* in C of the $\neg$ on the left of A.

If $(A * B)$ is a segment of C, then A and B are called the *left* and *right scopes* in C of the $*$ between A and B.

Note that A, B, and C in Definition 2.3.5 are formulas.

Theorem 2.3.6.

Any $\neg$ in any A has a unique scope. Any $*$ in any A has unique left and right scopes.

Proof. Any $\neg$ occurs in A by an application of the formation rule concerning $\neg$. Hence there is some B such that $(\neg B)$ is a segment of A. B is the scope of that $\neg$ in A.

Similarly for the left and right scopes of binary connectives.

We shall now prove the uniqueness of scopes. Consider any $\neg$ in A. Suppose both B and B' are its scopes in A. By Definition 2.3.5, both $(\neg B)$ and $(\neg B')$ are segments of A. Since the $\neg$'s on the left of B and B' are the same occurrence in A, by Lemma 2.3.2, $B = B'$. The scope of $\neg$ is thus unique.

Consider any $*$ in A. Suppose both C_1 and C_1' are its left scopes and both C_2 and C_2' are its right scopes in A. Then both $(C_1 * C_2)$ and $(C_1' * C_2')$ are segments of A. Since the $*$'s between C_1 and C_2 and between C_1' and C_2' are the same occurrence in A, C_1 and C_1' end with the same occurrence of a symbol of A, and C_2 and C_2' begin with the same occurrence of a symbol of A. By Lemma 2.3.2, $C_1 = C_1'$ and $C_2 = C_2'$. Thus, both the left and right scopes of $*$ are unique. □

Example

Suppose $A = (\neg(\,(p \wedge q\,) \vee (\,(\neg p) \to r\,)\,)\,)$. The scope of the first $\neg$ is $((p \wedge q) \vee ((\neg p) \to r))$; that of the second $\neg$ is p. The left and right scopes of $\wedge$ are p and q; those of $\vee$ are $(p \wedge q)$ and $((\neg p) \to r)$; those of $\to$ are $(\neg p)$ and r. We can verify the uniqueness of these scopes after reading this section.

Remarks

(1) In the part of proving (3) in the proof of Theorem 2.3.3, we derive $A = A_1$ directly from $(\neg A) = (\neg A_1)$, since the symbols $(, \neg,)$ on the left and right sides are respectively the same occurrences of symbols. But in the proof of Theorem 2.3.6, both $(\neg\, B)$ and $(\neg\, B')$ are segments of A, and the $\neg$'s on the left of B and B' are the same occurrence in A. Hence the ('s on the left of the two $\neg$'s are the same occurrence in A, but we cannot derive that the)'s on the right of B and B' are the same occurrence in A. Therefore we cannot derive $(\neg B) = (\neg B')$, nor derive $B = B'$ directly.

(2) By the uniqueness of the generation of formulas of $\mathcal{L}^p$ and the uniqueness of the scopes of connectives in formulas, it can easily be seen that, if A is a segment of B, then any connective of A has the same scope (or scopes) in A as in B.

Theorem 2.3.7.

[1] If A is a segment of $(\neg B)$, then A is a segment of B or $A = (\neg B)$.

[2] If A is a segment of $(B * C)$, then A is a segment of B or a segment of C or $A = (B * C)$.

Proof. In other words, [1] states that, if A is a proper segment of $(\neg B)$, then A is a segment of B. Now suppose A is a proper segment of $(\neg B)$.

If A contains the first symbol (of $(\neg B)$, then A is a proper initial segment of $(\neg B)$ and accordingly is not a formula. If A contains the last symbol), then A is a proper terminal segment of $(\neg B)$ and is also not a formula.

Suppose A contains the $\neg$ on the left of B. A must contain the initial (, otherwise A begins with $\neg$, which is impossible. Then A is a proper initial segment of $(\neg B)$ and is not a formula.

All the three cases contradict the well-formedness of A. Therefore, A contains none of the three symbols, that is, A is a segment of B.

Now we are to prove [2], which states that, if A is a proper segment of $(B * C)$, then A is a segment of B or C.

If A contains the first (or the last) of $(B * C)$, then A is not a formula (as seen in the proof of [1]), which yields a contradiction.

Suppose A contains the $*$ between B and C. Since A is a formula, this $*$ has scopes in A, say B_1 and C_1. By Definition 2.3.5, A contains $(B_1 * C_1)$ as its segment. The left and right scopes of this $*$ in $(B * C)$ are B and C. By the remark (2) before Theorem 2.3.7, we have $B = B_1$ and $C = C_1$, hence $(B * C) = (B_1 * C_1)$. Since $(B_1 * C_1)$ is a segment of A, and A is a segment of $(B * C)$, we have $(B * C) = A = (B_1 * C_1)$, contradicting the supposition that A is a proper segment of $(B * C)$.

Therefore, A contains none of the three symbols. That is, A is a segment of B or C. □

Example

Suppose A is a segment of $(\neg B)$:

$$(\neg B) = (\neg(p \wedge ((\neg q) \rightarrow r))).$$

A may be a segment of B, that is, A may be p, q, $(\neg q)$, r, $((\neg q) \rightarrow r)$,or B. However, if A contains any one of the three symbols of $(\neg B)$: the first symbol (, the last symbol), and the $\neg$ on the left of B, then necessarily $A = (\neg B)$. In other words, any proper segment of $(\neg B)$ containing one of these symbols is not a formula.

Suppose A is a segment of $(B * C)$:

$$(B * C) = ((\neg p) * (q \vee r)).$$

A may be a segment of B or C, that is, A may be p, $(\neg p)$, q, r, or $(q \vee r)$. However, if A contains any one of the three symbols of $(B * C)$: the first (, the last), and the $*$ between B and C, then necessarily $A = (B * C)$.

An *algorithm* is an effective procedure by means of which it can be decided (in a finite number of steps), for each member of a set, whether or not the member possesses some given property. One algorithm for deciding whether an expression is a formula of $\mathcal{L}^p$ is given as follows.

Let U be an expression of $\mathcal{L}^p$.

Step 1. An empty expression is not a formula.

Step 2. A single symbol is a formula iff it is a proposition symbol.

Step 3. U with more than one symbol must begin with a left parenthesis, otherwise U is not a formula. If the second symbol is $\neg$, U must be $(\neg V)$, where V is an expression, otherwise U is not a formula. Then the question of well-formedness of U is reduced to the same question on a shorter expression V. Return to step 1.

Step 4. If U begins with a left parenthesis but its second symbol is not $\neg$, scan U from left to right until reaching (V, where V is an expression with the same number of occurrences of left and right parentheses. (If the end of U is reached before meeting such a V, then U is not a formula.) U must be $(V * W)$, where W is an expression, otherwise U is not a formula. Then the question of well-formedness of U is reduced to the same question on shorter expressions V and W. Return to Step 1.

Because every expression is finite in length, the above procedure terminates after finite steps. It is left to the reader to verify that the above steps do constitute an algorithm for deciding whether a given expression is a formula of $\mathcal{L}^p$.

In this and the last sections, formulas are defined and their structural properties are discussed. These belong to the syntax of a formal language, because they are not concerned with the meaning of symbols and formulas.

Before finishing this section, we want to introduce some conventions for omitting the parentheses in formulas to facilitate reading.

The outermost parentheses are usually omitted. For instance,

$$((p \wedge q) \to (p \vee r))$$

is usually written as

$$(p \wedge q) \to (p \vee r).$$

Parentheses may be used together with brackets and curly brackets:

$$(\ )\ [\ ]\ \{\ \}$$

Thus

$$((p \wedge q) \to (p \vee r)) \leftrightarrow (\neg q)$$

may be written more clearly as

$$[(p \wedge q) \to (p \vee r)] \leftrightarrow (\neg q).$$

We also often omit the parentheses, subject to the convention of priority. In algebra $x + y \cdot z$ means $x + (y \cdot z)$. It is said that $\cdot$ has priority over $+$. In the following sequence:

$$\neg \quad \wedge \quad \vee \quad \rightarrow \quad \leftrightarrow$$

each connective on the left has priority over those on the right. Accordingly the formula

1) $$(((\neg p) \rightarrow ((p \wedge (\neg q)) \vee r)) \leftrightarrow q)$$

may be written as

2) $$\neg p \rightarrow p \wedge \neg q \vee r \leftrightarrow q$$

with all its parentheses omitted.

We shall not always omit the maximum number of parentheses which our conventions would allow, but aim at securing maximum readability. Hence it is perhaps more profitable to write 1) as

$$[\neg p \rightarrow (p \wedge \neg q) \vee r] \leftrightarrow q.$$

It has been mentioned above that the aim of omitting parentheses in formulas is to facilitate reading. When we want to decide whether an expression is a formula, we should write it in its original (unabbreviated) form with no parentheses omitted.

Exercises 2.3.

2.3.1. Show that no formula of $\mathcal{L}^p$ is of length 2, 3, or 6, but any other length is possible.

2.3.2. Suppose U, V and W are non-empty expressions of $\mathcal{L}^p$. Show that at most one of UV and VW is a formula.

2.4. SEMANTICS

In this section we want to explain how to interpret the propositional language $\mathcal{L}^p$ and make the formulas express propositions. We first give some intuitive illustrations.

Formulas are composed of atoms (proposition symbols) and connectives. Atoms are intended to express simple propositions. The connectives have their intended meanings: negation, conjunction, disjunction, implication, and equivalence express, respectively, "not", "and", "(inclusive) or", "if, then", and "iff". Hence, if formulas A and B express propositions $\mathcal{A}$ and $\mathcal{B}$ respectively, then the following non-atomic formulas on the left express the corresponding compound propositions on the right:

$\neg$A	Not $\mathcal{A}$.
A $\wedge$ B	$\mathcal{A}$ and $\mathcal{B}$.
A $\vee$ B	$\mathcal{A}$ or $\mathcal{B}$.
A $\rightarrow$ B	If $\mathcal{A}$ then $\mathcal{B}$.
A $\leftrightarrow$ B	$\mathcal{A}$ iff $\mathcal{B}$.

We need not know what propositions $\mathcal{A}$ and $\mathcal{B}$ are, because, as mentioned in the Introduction, deducibility is not concerned with the matter of premises and conclusions.

Formulas have no values but we may assign to them the values of the propositions they express. Then the value of $\neg$A is determined by that of A, and the values of A $\wedge$ B, A $\vee$ B, A $\rightarrow$ B, and A $\leftrightarrow$ B are determined by those of A and B, as shown in the following tables:

A	$\neg$ A
1	0
0	1

A	B	A $\wedge$ B	A $\vee$ B	A $\rightarrow$ B	A $\leftrightarrow$ B
1	1	1	1	1	1
1	0	0	1	0	0
0	1	0	1	1	0
0	0	0	0	1	1

These tables are called the *truth tables* of negation, conjuction, disjunction, implication, and equivalence.

The above illustrations lead us to the following definitions.

Definition 2.4.1. (***Truth valuation***)

A *truth valuation* is a function with the set of all proposition symbols as domain and $\{1, 0\}$ as range.

By Definition 2.4.1, a truth valuation assigns a value to every proposition symbol simultaneously. We use the italic small Latin letter t (with or without subscripts or superscripts) to denote any truth valuation. The value which t assigns to any formula A (to be defined below) is written as A^t.

Definition 2.4.2. (***Values of formulas***)

The *value* assigned to formulas by a truth valuation t is defined by recursion:

[1] $p^t \in \{1, 0\}$.

[2] $(\neg A)^t = \begin{cases} 1 & \text{if } A^t = 0, \\ 0 & \text{otherwise.} \end{cases}$

[3] $(A \wedge B)^t = \begin{cases} 1 & \text{if } A^t = B^t = 1, \\ 0 & \text{otherwise.} \end{cases}$

[4] $(A \vee B)^t = \begin{cases} 1 & \text{if } A^t = 1 \text{ or } B^t = 1, \\ 0 & \text{otherwise.} \end{cases}$

[5] $(A \to B)^t = \begin{cases} 1 & \text{if } A^t = 0 \text{ or } B^t = 1, \\ 0 & \text{otherwise.} \end{cases}$

[6] $(A \leftrightarrow B)^t = \begin{cases} 1 & \text{if } A^t = B^t, \\ 0 & \text{otherwise.} \end{cases}$

Theorem 2.4.3.

For any $A \in \mathit{Form}(\mathcal{L}^p)$ and any truth valuation$t$, $A^t \in \{1, 0\}$.

Proof. By induction on the structure of A. □

A truth valuation assigns a value to each proposition symbol. But the value A^t, which t assigns to a formula A, is concerned only with the values which t assigns to the proposition symbols occurring in A.

Example

Suppose $A = p \vee q \to q \wedge r$ and t is a truth valuation such that

$$p^t = q^t = r^t = 1. \tag{1}$$

Then we have

$$\begin{aligned}(\mathrm{p} \vee \mathrm{q})^t &= 1,\\ (\mathrm{q} \wedge \mathrm{r})^t &= 1,\\ \mathrm{A}^t &= 1.\end{aligned}$$

Suppose t_1 is another truth valuation such that

(2) $$\mathrm{p}^{t_1} = \mathrm{q}^{t_1} = \mathrm{r}^{t_1} = 0.$$

Then we have

$$\begin{aligned}(\mathrm{p} \vee \mathrm{q})^{t_1} &= 0,\\ (\mathrm{q} \wedge \mathrm{r})^{t_1} &= 0,\\ \mathrm{A}^{t_1} &= 1.\end{aligned}$$

In fact, $\mathrm{p}^t = 1$ in (1) is unnecessary, because $\mathrm{q}^t = 1$ and $\mathrm{r}^t = 1$ are sufficient for $\mathrm{A}^t = 1$. $\mathrm{r}^{t_1} = 0$ in (2) is also unnecessary.

If t_2 is a third truth valuation such that $\mathrm{p}^{t_2} = 1$ and $\mathrm{r}^{t_2} = 0$, then $\mathrm{A}^{t_2} = 0$.

The above example illustrates that the values which various truth valuations assign to a formula may or may not be different.

We use the capital roman-type Greek letter Σ (with or without subscripts or superscripts) to denote any set of formulas. That is, Σ denotes all the formulas in it. We define

$$\Sigma^t = \begin{cases} 1 & \text{if for each } \mathrm{B} \in \Sigma,\ \mathrm{B}^t = 1,\\ 0 & \text{otherwise.}\end{cases}$$

Note that $\Sigma^t = 0$ means that exists $\mathrm{B} \in \Sigma$ such that $\mathrm{B}^t = 0$, not that for all $\mathrm{B} \in \Sigma$, $\mathrm{B}^t = 0$.

Definition 2.4.4. (***Satisfiability***)
Σ is *satisfiable* iff there is some truth valuation t such that $\Sigma^t = 1$.
When $\Sigma^t = 1$, t is said to *satisfy* Σ.

Obviously, the satisfiability of Σ implies that all the formulas in it will be satisfiable. But the converse of this implication does not hold, because the satisfiability of Σ requires a single truth valuation satisfying all the formulas in it.

Definition 2.4.5. (***Tautology*, *Contradiction***)
A is a *tautology* iff, for any truth valuation t, $A^t = 1$.
A is a *contradiction* iff, for any truth valuation t, $A^t = 0$.

By definition, the question whether a formula is a tautology or a contradiction or neither, is concerned with all the truth valuations which are infinite in number. (A formula is satisfiable iff its negation is not a tautology.) However, as mentioned before, the value A^t is concerned only with the values which t assigns to the different proposition symbols $p_1, \ldots, p_n$ occurring in A. There are 2^n different truth valuations for $p_1, \ldots, p_n$. In each of the 2^n cases the value of A is obtained effectively. The values of A in all the 2^n cases form a table, called the *truth table* of A. It can be seen in this table whether A is a tautology or a contradiction or neither.

Example
The truth table of $(p \vee q \to r) \leftrightarrow (p \to r) \wedge (q \to r)$ is

p q r	(p	$\vee$	q	$\to$	r)	$\leftrightarrow$	(p	$\to$	r)	$\wedge$	(q	$\to$	r)
1 1 1	1	1	1	1	1	1	1	1	1	1	1	1	1
1 1 0	1	1	1	0	0	1	1	0	0	0	1	0	0
1 0 1	1	1	0	1	1	1	1	1	1	1	0	1	1
1 0 0	1	1	0	0	0	1	1	0	0	0	0	1	0
0 1 1	0	1	1	1	1	1	0	1	1	1	1	1	1
0 1 0	0	1	1	0	0	1	0	1	0	0	1	0	0
0 0 1	0	0	0	1	1	1	0	1	1	1	0	1	1
0 0 0	0	0	0	1	0	1	0	1	0	1	0	1	0

The table is formed as follows. First the formula is written; and also its three atoms p, q, and r (on its left). The $2^3 = 8$ different truth valuations for p, q, r are written below them. For instance, in the third case: 1,0,1 is assigned to p, q, r respectively. In each case, the values of the atoms are copied below each occurrence of them in the formula. Then the values of the various well-formed segments of the entire formula are obtained in order of increasing length of the segments; the values of each segment being written below the connective used in forming it. For instance, the values of $(p \to r)$ are written below the $\to$, which is used in forming it. Finally, the values of the entire formula in each case are written in a column below $\leftrightarrow$, which is used in the last step of its formation.

From the above example, it is easily seen that a formula containing three different proposition symbols corresponds to a ternary truth function. In general, a formula containing n different proposition symbols corresponds to an n-ary truth function.

The above procedure is generally applicable, but another one to be introduced is perhaps more efficient.

The "expressions" A ∧ 1, 0 ∨ A, etc. in the following table are not expressions, because 1 and 0 are not symbols of the formal language. In fact, the A occurring in them does not denote a formula, but is intended to denote the value of a formula. They are used here for the provisional purpose of evaluating formulas.

$\neg 1$	0
$\neg 0$	1
$A \wedge 1$	A
$1 \wedge A$	A
$A \wedge 0$	0
$0 \wedge A$	0
$A \vee 1$	1
$1 \vee A$	1
$A \vee 0$	A
$0 \vee A$	A
$A \rightarrow 1$	1
$1 \rightarrow A$	A
$A \rightarrow 0$	$\neg A$
$0 \rightarrow A$	1
$A \leftrightarrow 1$	A
$1 \leftrightarrow A$	A
$A \leftrightarrow 0$	$\neg A$
$0 \leftrightarrow A$	$\neg A$

Such "expressions" in the above table have the same value as those on the right. However, those on the right are simpler. Therefore we may replace the "expressions" on the left by the corresponding ones on the right to simplify the evaluation.

Now we are in a position to describe the procedure. Suppose a formula A is given. Assign 1 to one of the atoms, say p, occurring in A, written as $p = 1$. Evaluate A by applying all possible simplifications given in the above table. The value of A may be 1 or 0 or may equal that of a new formula with no p occurring in it. Then set $p = 0$ and make the evaluation of A. Thus, we begin with A and obtain two branches by setting $p = 1$ and $p = 0$, each yielding a new formula or a value. A value is called a terminal. For a non-terminal new formula, we continue the branching process described above. The process terminates when values (terminals) occur in all the branches. The given formula is a tautology if all the terminals are 1, or a contradiction if all the terminals are 0, or neither.

Example

Let $A = (p \wedge q \to r) \wedge (p \to q) \to (p \to r)$. Then we have

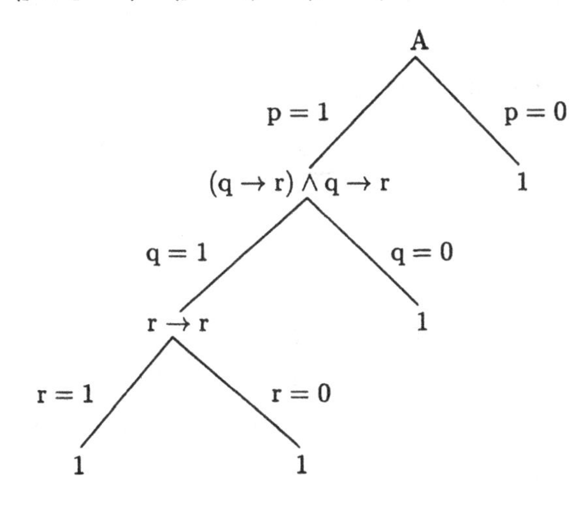

Therefore A is a tautology.

Exercises 2.4.

2.4.1. For each of the following formulas decide whether it is a tautology or a contradiction or neither:

[1] $(p \vee q \to r) \to (p \to q \wedge r)$

[2] $(p \wedge \neg q) \wedge (\neg p \vee q)$

[3] $(p \to q) \leftrightarrow (q \to p)$
[4] $(p \to r) \wedge (q \to r) \leftrightarrow (p \vee q \to r)$
[5] $(p \to r) \vee (q \to r) \leftrightarrow (p \wedge q \to r)$

2.4.2. Suppose $A = p_1 \leftrightarrow (\ldots(p_{n-1} \leftrightarrow p_n)\ldots)$ and t is a truth valuation. Prove that $A^t = 1$ iff $p_i^t = 0$ for an even number of i's ($i = 1, \ldots, n$). (*Hint.* Use the tautologies

$$(p \leftrightarrow q) \leftrightarrow (q \leftrightarrow p),$$
$$[(p \leftrightarrow q) \leftrightarrow r] \leftrightarrow [p \leftrightarrow (q \leftrightarrow r)]$$

or by induction on n. The two tautologies mean that $\leftrightarrow$ satisfies commutative and associative laws, hence any permutation of the proposition symbols in A does not change the truth value of A.)

2.4.3. Suppose A is a formula containing only the connective $\leftrightarrow$. Prove that A is a tautology iff each atom occurs an even number of times in A. (*Hint.* Use the tautologies given in the above hint and the fact that if B is a tautology then $B \leftrightarrow C$ is a tautology iff C is a tautology.)

2.4.4. For what n is the formula

$$\underbrace{(\ldots((A \to A) \to A)\ldots) \to A}_{n}$$

a tautology?

2.4.5. Suppose

$$(A_i \to B_i)^t = 1 \ \ (i = 1, \ldots, n),$$
$$(A_1 \vee \ldots \vee A_n)^t = 1,$$
$$(B_i \wedge B_j)^t = 0 \ \text{ for } i \neq j \ \ (i, j = 1, \ldots, n).$$

Prove that $(B_i \to A_i)^t = 1 \ \ (i = 1, \ldots, n)$.

2.5. TAUTOLOGICAL CONSEQUENCE

Suppose $\mathcal{A}_1, \ldots, \mathcal{A}_n$, and $\mathcal{A}$ are propositions. Deductive logic studies whether

$$\mathcal{A} \text{ is deducible from } \mathcal{A}_1, \ldots, \mathcal{A}_n.$$

(that is, the truth of $\mathcal{A}_1, \ldots, \mathcal{A}_n$ implies that of $\mathcal{A}$) holds. Suppose $\mathcal{A}_1, \ldots,$ $\mathcal{A}_n$ and $\mathcal{A}$ are expressed respectively by formulas $\mathrm{A}_1, \ldots, \mathrm{A}_n$ and A. Then the question raised is: what kind of relation between $\{\mathrm{A}_1, \ldots, \mathrm{A}_n\}$ and A corresponds to the deducibility relation between $\{\mathcal{A}_1, \ldots, \mathcal{A}_n\}$ and $\mathcal{A}$?

Formulas are assigned values by truth valuations. This makes us consider the following relation between $\{\mathrm{A}_1, \ldots, \mathrm{A}_n\}$ and A: if a truth valuation assigns truth to each of $\mathrm{A}_1, \ldots, \mathrm{A}_n$, then it assigns truth to A. However, since different truth valuations may assign different values to the same formula, this relation should be stated more precisely by requiring the truth valuation to be arbitrary. This leads us to the following definition.

Definition 2.5.1. (***Tautological consequence***)
Suppose $\Sigma \subseteq Form(\mathcal{L}^p)$ and $\mathrm{A} \in Form(\mathcal{L}^p)$.

A is a *tautological consequence* of Σ (that is, of the formulas in Σ), written as $\Sigma \models \mathrm{A}$, iff for any truth valuation t, $\Sigma^t = 1$ implies $\mathrm{A}^t = 1$.

Note that the notation $\models$ is not a symbol of the formal language and hence $\Sigma \models \mathrm{A}$ is not a formula. $\Sigma \models \mathrm{A}$ is a proposition (in the metalanguage) about Σ and A. $\models$ may be read as "logically implies".

We write $\Sigma \not\models \mathrm{A}$ for "not $\Sigma \models \mathrm{A}$". That is, there exists some truth valuation t such that $\Sigma^t = 1$ and $\mathrm{A}^t = 0$.

Σ is any set of formulas. When Σ is the empty set, we obtain the important special case $\emptyset \models \mathrm{A}$ of tautological consequences. By Definition 2.5.1, $\emptyset \models \mathrm{A}$ is

1) For any t, if $\emptyset^t = 1$, then $\mathrm{A}^t = 1$.

where $\emptyset^t = 1$ means

2) For any B, if $\mathrm{B} \in \emptyset$, then $\mathrm{B}^t = 1$.

Because $\mathrm{B} \in \emptyset$ is false, 2) is vacuously true. Hence "if $\emptyset^t = 1$, then $\mathrm{A}^t = 1$" in 1) is equivalent to $\mathrm{A}^t = 1$. Thus $\emptyset \models \mathrm{A}$ means that A is a tautology.

Intuitively speaking, $\Sigma \models \mathrm{A}$ means that the truth of the formulas in Σ is the sufficient condition of the truth of A. Since $\emptyset$ consists of no formula, $\emptyset \models \mathrm{A}$ means that the truth of A is unconditional, hence A is a tautology.

For two formulas, we write

$$\mathrm{A} \models\!\!\!\dashv \mathrm{B}$$

to denote "$\mathrm{A} \models \mathrm{B}$ and $\mathrm{B} \models \mathrm{A}$". A and B are said to be *tautologically equivalent* (or simply, *equivalent*) iff $\mathrm{A} \models\!\!\!\dashv \mathrm{B}$ holds. Tautologically equivalent formulas are assigned the same value by any truth valuation.

We now show how to prove or refute a tautological consequence. In general, to prove $\Sigma \models A$ we must show that any truth valuation satisfying Σ satisfies A. To refute $\Sigma \models A$ (that is, to prove $\Sigma \not\models A$) we must construct a truth valuation satisfying Σ but not satisfying A.

Example
$A \to B,\ \ B \to C \models A \to C.$

Proof. Suppose $A \to B$, $B \to C \not\models A \to C$. That is, there is a truth valuation t such that

$$(A \to B)^t = 1, \tag{1}$$
$$(B \to C)^t = 1, \tag{2}$$
$$(A \to C)^t = 0. \tag{3}$$

By (3) we have

$$A^t = 1, \tag{4}$$
$$C^t = 0. \tag{5}$$

By (1) and (4) we have $B^t = 1$; and from $B^t = 1$ and (2) we have $C^t = 1$, which contradicts (5). Hence the tautological consequence is proved.

Example
$(A \to \neg B) \vee C,\ \ B \wedge \neg C,\ \ A \leftrightarrow C \not\models \neg A \wedge (B \to C).$

Proof. Let t be a truth valuation such that $A^t = 0$, $B^t = 1$, and $C^t = 0$. Then we have

$$((A \to \neg B) \vee C)^t = 1,$$
$$(B \wedge \neg C)^t = 1,$$
$$(A \leftrightarrow C)^t = 1,$$
$$(\neg A \wedge (B \to C))^t = 0,$$

which proves the statement.

Remarks
In the first example above, we begin with (3) because we can deduce (4) and (5) from it. Thus, the tautological consequence is easily proved. If we begin with (1), and deduce from it that "$A^t = 1$ and $B^t = 1$" or "$A^t = 0$ and $B^t = 1$" or "$A^t = 0$ and $B^t = 0$", then the proof is more complicated.

Similarly in the second example, it is convenient to first make t satisfy $(B \wedge \neg C)^t = 1$, from which we get $B^t = 1$ and $C^t = 0$, then $A^t = 0$ is obtained from $(A \leftrightarrow C)^t = 1$.

Conjunction and disjunction satisfy both the commutative and associative laws (the proofs are immediate):

$$A \wedge B \models\!\!\!\dashv B \wedge A,$$
$$(A \wedge B) \wedge C \models\!\!\!\dashv A \wedge (B \wedge C),$$
$$A \vee B \models\!\!\!\dashv B \vee A,$$
$$(A \vee B) \vee C \models\!\!\!\dashv A \vee (B \vee C).$$

These laws also hold in formal deducibility (see Theorems 2.6.8 and 2.6.9 in the next section):

$$A \wedge B \vdash\!\dashv B \wedge A,$$
$$(A \wedge B) \wedge C \vdash\!\dashv A \wedge (B \wedge C),$$
$$A \vee B \vdash\!\dashv B \vee A,$$
$$(A \vee B) \vee C \vdash\!\dashv A \vee (B \vee C).$$

Thus, we may write

$$A_1 \wedge \ldots \wedge A_n$$
$$A_1 \vee \ldots \vee A_n$$

without parentheses and alter the order of the conjuncts and disjuncts.

Theorem 2.5.2.
[1] $A_1, \ldots, A_n \models A$ iff $\emptyset \models A_1 \wedge \ldots \wedge A_n \to A$.
[2] $A_1, \ldots, A_n \models A$ iff $\emptyset \models A_1 \to (\ldots (A_n \to A) \ldots)$. □

Lemma 2.5.3.
If $A \models\!\!\!\dashv A'$ and $B \models\!\!\!\dashv B'$, then
[1] $\neg A \models\!\!\!\dashv \neg A'$.
[2] $A \wedge B \models\!\!\!\dashv A' \wedge B'$.
[3] $A \vee B \models\!\!\!\dashv A' \vee B'$.
[4] $A \to B \models\!\!\!\dashv A' \to B'$.
[5] $A \leftrightarrow B \models\!\!\!\dashv A' \leftrightarrow B'$. □

Theorem 2.5.4. (***Replaceability of equivalent formulas***)

If B $\models\!\mid$ C and A$'$ result from A by replacing some (not necessarily all) occurrences of B in A by C, then A $\models\!\mid$ A$'$.

Proof. By induction on the structure of A.

If B = A, then C = A$'$. This theorem thus holds.

Basis. A is an atom. Then B = A; the theorem holds.

Induction step. A is one of the five forms: $\neg A_1$, $A_1 \wedge A_2$, $A_1 \vee A_2$, $A_1 \to A_2$, or $A_1 \leftrightarrow A_2$.

Suppose A = $\neg A_1$. If B = A, the theorem holds as stated above. If B $\neq$ A, then B is a segment of A_1 (by Theorem 2.3.7). Let A_1' results from A_1 by the replacement stated in the theorem, then $A' = \neg A_1'$. We have

$$A_1 \models\!\mid A_1' \quad \text{(by ind hyp)},$$
$$\neg A_1 \models\!\mid \neg A_1' \quad \text{(by Lem 2.5.3 [1])}.$$

That is, A $\models\!\mid$ A$'$.

Suppose A = $A_1 * A_2$. ($*$ denotes any one of $\wedge$, $\vee$, $\to$, $\leftrightarrow$.) If B = A, the theorem holds as in the above case. If B $\neq$ A, then B is a segment of A_1 or A_2 (by Theorem 2.3.7). Let A_1' and A_2' result respectively from A_1 and A_2 by the replacement stated in the theorem, then $A' = A_1' * A_2'$. We have

$$A_1 \models\!\mid A_1',\ A_2 \models\!\mid A_2' \quad \text{(by ind hyp)},$$
$$A_1 * A_2 \models\!\mid A_1' * A_2' \quad \text{(by Lem 2.5.3 [2]–[5])}.$$

That is, A $\models\!\mid$ A$'$.

By the basis and induction step, the theorem is proved. □

Theorem 2.5.5. (***Duality***)

Suppose A is a formula composed of atoms and the connectives $\neg$, $\wedge$, and $\vee$ by the formation rules concerned, and A$'$ results by exchanging in A $\wedge$ for $\vee$ and each atom for its negation. Then A$'$ $\models\!\mid$ $\neg$A. (A$'$ is the *dual* of A.)

Proof. By induction on the structure of A. □

Exercises 2.5.

2.5.1. Prove Theorem 2.5.5.

2.5.2. Prove the following:

[1] $\neg(A \wedge B) \models\!\!\!\!\!\dashv \neg A \vee \neg B$.

[2] $\neg(A \vee B) \models\!\!\!\!\!\dashv \neg A \wedge \neg B$.

[3] $A \to (B \wedge C) \models\!\!\!\!\!\dashv (A \to B) \wedge (A \to C)$.

[4] $A \to (B \vee C) \models\!\!\!\!\!\dashv (A \to B) \vee (A \to C)$.

[5] $(A \wedge B) \to C \models\!\!\!\!\!\dashv (A \to C) \vee (B \to C)$.

[6] $(A \vee B) \to C \models\!\!\!\!\!\dashv (A \to C) \wedge (B \to C)$.

[7] $A \to (B \to C) \models B \leftrightarrow B \wedge (A \leftrightarrow A \wedge C)$.

2.5.3. Prove the following:

[1] $(A \to B) \vee (A \to C) \not\models A \to (B \wedge C)$.

[2] $A \to (B \vee C) \not\models (A \to B) \wedge (A \to C)$.

[3] $(A \wedge B) \to C \not\models (A \to C) \wedge (B \to C)$.

[4] $(A \to C) \vee (B \to C) \not\models (A \vee B) \to C$.

2.6. FORMAL DEDUCTION

We have mentioned in the Introduction that Leibniz looked for a calculus of reasoning. It is the formal deduction to be formulated in this section. The correctness of formal deduction can be checked mechanically.

We proved in the last section

1) $$A \to B, B \to C \models A \to C.$$

Suppose A, B and C express respectively the propositions $\mathcal{A}$, $\mathcal{B}$ and $\mathcal{C}$. Then 1) corresponds to

2) From "if $\mathcal{A}$ then $\mathcal{B}$" and "if $\mathcal{B}$ then $\mathcal{C}$" we deduce "if $\mathcal{A}$ then $\mathcal{C}$".

2) is proved as follows. Suppose 2) does not hold. That is, both "if $\mathcal{A}$ then $\mathcal{B}$" and "if $\mathcal{B}$ then $\mathcal{C}$" are true, but "if $\mathcal{A}$ then $\mathcal{C}$" is false. Then, $\mathcal{A}$ is true and $\mathcal{C}$ is false. Consequently, $\mathcal{B}$ is true and $\mathcal{C}$ is true too, thus yielding a contradiction. Hence 2) holds.

1) corresponds to 2). The proof of 1) in the last section is analogous to that of 2). The distinction between them lies in that 1) is concerned with formulas, while 2) is concerned with propositions. Undoubtedly, either of the proofs is correct. However, we cannot mechanically check their correctness, and cannot even check whether they are proofs, because we have not defined the concept of a proof.

We want to define another kind of relation, called formal deducibility. The significance of the word "formal" will be explained later. The important point is that formal deducibility is concerned with the syntactic structure of formulas and its proof can be checked mechanically.

First of all some notational conventions will be introduced.

Suppose $\Sigma = \{A_1, A_2, A_3, \ldots\}$. For convenience, Σ may be written as a sequence A_1, A_2, $A_3, \ldots$. Written in this way, however, the order of the members is irrelevant, because Σ is a set. Accordingly, the sets $\Sigma \cup \{A\}$ and $\Sigma \cup \Sigma'$ may be written as Σ, A and Σ, Σ', respectively.

We use the symbol $\vdash$ to denote the relation of formal deducibility and write

$$\Sigma \vdash A$$

to mean that A is formally deducible (or provable) from Σ. Formal deducibility is a relation between Σ (a set of formulas which are the premises) and A (a formula which is the conclusion). Note that $\vdash$ is not a symbol of the formal language and $\Sigma \vdash A$ is not a formula. $\Sigma \vdash A$ is a proposition (in the metalanguage) about Σ and A. "$\vdash$" may be read as "yields".

Formal deducibility will be defined by rules of formal deduction. In propositional logic there are eleven rules of formal deduction formulated as follows.

(Ref) $A \vdash A$. (*Reflexivity*)

(+) If $\Sigma \vdash A$,
then $\Sigma, \Sigma' \vdash A$. (*Addition of premises*)

($\neg-$) If $\Sigma, \neg A \vdash B$,
$\Sigma, \neg A \vdash \neg B$,
then $\Sigma \vdash A$. ($\neg$*-elimination*)

($\to-$) If $\Sigma \vdash A \to B$,
$\Sigma \vdash A$,
then $\Sigma \vdash B$. ($\to$*-elimination*)

($\to+$) If $\Sigma, A \vdash B$,
then $\Sigma \vdash A \to B$. ($\to$*-introduction*)

($\wedge-$) If $\Sigma \vdash A \wedge B$,
then $\Sigma \vdash A$,

$\Sigma \vdash B$. ($\wedge$-*elimination*)

($\wedge$+) If $\Sigma \vdash A$,
$\Sigma \vdash B$,
then $\Sigma \vdash A \wedge B$. ($\wedge$-*introduction*)

($\vee$−) If $\Sigma, A \vdash C$,
$\Sigma, B \vdash C$,
then $\Sigma, A \vee B \vdash C$. ($\vee$-*elimination*)

($\vee$+) If $\Sigma \vdash A$,
then $\Sigma \vdash A \vee B$,
$\Sigma \vdash B \vee A$. ($\vee$-*introduction*)

($\leftrightarrow$−) If $\Sigma \vdash A \leftrightarrow B$,
$\Sigma \vdash A$,
then $\Sigma \vdash B$.
If $\Sigma \vdash A \leftrightarrow B$,
$\Sigma \vdash B$,
then $\Sigma \vdash A$. ($\leftrightarrow$-*elimination*)

($\leftrightarrow$+) If $\Sigma, A \vdash B$,
$\Sigma, B \vdash A$,
then $\Sigma \vdash A \leftrightarrow B$. ($\leftrightarrow$-*introduction*)

Each of these rules is not a single rule, but a scheme of rules, because Σ is any set of formulas, and A, B, and C are any formulas. Examples are first given to explain how the rules are applied.

Example Suppose $A \in \Sigma$ and $\Sigma' = \Sigma - \{A\}$. The following sequence:

(1) $A \vdash A$ (by (Ref)).
(2) $A, \Sigma' \vdash A$ (by $(+), (1)$).
(That is, $\Sigma \vdash A$.)

consists of two steps. Step (1) is generated directly by the rule (Ref). Step (2) is generated by the rule (+), which is applied to step (1). At each of the steps, the rule applied and the preceding steps concerned (if any) form a justification for this step, and are written on the right. These steps are said to form a proof of the last step.

Hence, it is proved in this example that when $A \in \Sigma$, $\Sigma \vdash A$ holds. It is denoted by (ϵ), using the notation for membership. It contains (Ref) as a special case.

Example

The following sequence

(1) $A \to B, B \to C, A \vdash A \to B$ (by (ϵ)).
(2) $A \to B, B \to C, A \vdash A$ (by (ϵ)).
(3) $A \to B, B \to C, A \vdash B$ (by $(\to -)$, (1), (2)).
(4) $A \to B, B \to C, A \vdash B \to C$ (by (ϵ)).
(5) $A \to B, B \to C, A \vdash C$ (by $(\to -)$, (4), (3)).
(6) $A \to B, B \to C \vdash A \to C$ (by $(\to +)$, (5)).

consists of six steps. At each step, one of the eleven rules or (ϵ), which has just been proved, is applied. On the right are written the justifications for the steps. These steps form a proof of

$$A \to B, \quad B \to C \vdash A \to C$$

which is generated in the last step.

A demonstrated $\Sigma \vdash A$ may be called a *scheme of formal deducibility.*

Among the eleven rules stated above, (Ref) is the only one which generates schemes of formal deducibility directly. No preceding step is concerned in the application of (Ref). One step is concerned in the application of the rules $(+)$, $(\to +)$, $(\wedge -)$, and $(\vee +)$; and two steps are concerned in the application of $(\neg -)$, $(\to -)$, $(\wedge +)$, $(\vee -)$, $(\leftrightarrow -)$, and $(\leftrightarrow +)$.

(ϵ) also generates schemes of formal deducibility directly.

Rules of formal deduction are only concerned with the syntactic structures of formulas. For instance, from

3) $$\Sigma, \neg A \vdash B$$
4) $$\Sigma, \neg A \vdash \neg B$$

we can generate

5) $$\Sigma \vdash A$$

by applying $(\neg -)$. The premise, Σ of 5), is the Σ in the premises of 3) and 4). The conclusion, A of 5), results by deleting the leftmost $\neg$ of $\neg A$ in the premises of 3) and 4). The B of 3) and 4) is an arbitrary formula. Therefore, it can be checked mechanically whether the rules are used correctly.

The elimination (introduction) of a connective means that one occurrence of this connective is eliminated (introduced) in the conclusion of the scheme of formal deducibility generated by the rule. For instance, in $(\to -)$:

$$\begin{array}{ll} \text{If} & \Sigma \vdash A \to B, \\ & \Sigma \vdash A, \\ \text{then} & \Sigma \vdash B. \end{array}$$

the $\to$ between A and B in $A \to B$ is eliminated in the conclusion B of $\Sigma \vdash B$, which is generated by this rule. In $(\to +)$:

$$\begin{array}{ll} \text{If} & \Sigma, A \vdash B, \\ \text{then} & \Sigma \vdash A \to B. \end{array}$$

the $\to$ between A and B in the conclusion $A \to B$ is introduced.

It should be pointed out that in $(\vee -)$ it is the $\vee$ between A and B in $A \vee B$ that is eliminated in the conclusion C.

The intuitive meanings of most of the rules are quite obvious, but those of $(\neg -)$, $(\vee -)$, and $(\to +)$ call for some explanations. $(\neg -)$ expresses the method of *indirect proof* in informal reasoning: if a contradiction (denoted by B and $\neg$B) follows from certain premises (denoted by Σ) with an additional supposition that a certain proposition does not hold (denoted by $\neg$A), then this proposition is deducible from the premises (denoted by $\Sigma \vdash A$). $(\vee -)$ expresses the method of *proof by cases*. If proposition $\mathcal{A}$ follows from $\mathcal{B}$ and $\mathcal{C}$ separately, then $\mathcal{A}$ follows from "$\mathcal{B}$ or $\mathcal{C}$".

$(\to +)$ expresses that to prove an implicational proposition "if $\mathcal{A}$ then $\mathcal{B}$" from certain premises (denoted by $\Sigma \vdash A \to B$), it is sufficient to prove $\mathcal{B}$ from the premises together with $\mathcal{A}$ (denoted by Σ, $A \vdash B$).

Then we can see how the proof of $A \to B$, $B \to C \vdash A \to C$ in the foregoing example expresses a proof in the informal reasoning: from "$\mathcal{A}$ implies $\mathcal{B}$", "$\mathcal{B}$ implies $\mathcal{C}$" and $\mathcal{A}$, we obtain $\mathcal{B}$; from "$\mathcal{B}$ implies $\mathcal{C}$" and $\mathcal{B}$ we obtain $\mathcal{C}$; hence $\mathcal{C}$ is obtained from "$\mathcal{A}$ implies $\mathcal{B}$", "$\mathcal{B}$ implies $\mathcal{C}$" and $\mathcal{A}$, and accordingly "$\mathcal{A}$ implies $\mathcal{C}$" is obtained from "$\mathcal{A}$ implies $\mathcal{B}$" and "$\mathcal{B}$ implies $\mathcal{C}$".

Now we state the definition of formal deducibility.

Definition 2.6.1. (***Formal deducibility***)

A is *formally deducible* from Σ, written as $\Sigma \vdash A$, iff $\Sigma \vdash A$ is generated by (a finite number of applications of) the rules of formal deduction.

By the above definition, $\Sigma \vdash A$ holds iff there is a finite sequence

$$\Sigma_1 \vdash A_1, \ldots, \Sigma_n \vdash A_n \tag{6)}$$

such that each term $\Sigma_k \vdash A_k$ ($k = 1, \ldots, n$) in 6) is generated by one rule of formal deduction and $\Sigma_n \vdash A_n$ is $\Sigma \vdash A$ (that is, $\Sigma_n = \Sigma$ and $A_n = A$).

To say $\Sigma_k \vdash A_k$ is generated by one rule of formal deduction, say $(\neg-)$, means that in the subsequence

$$\Sigma_1 \vdash A_1, \ldots, \Sigma_{k-1} \vdash A_{k-1} \tag{7)}$$

which precedes $\Sigma_k \vdash A_k$ in 6), there are two terms

$$\Sigma_k, \neg A_k \vdash B,$$
$$\Sigma_k, \neg A_k \vdash \neg B,$$

where B is an arbitrary formula. In another example, if $\Sigma_k \vdash A_k$ is generated by $(\vee-)$, then there are in 7) two terms

$$\Sigma', B \vdash A_k,$$
$$\Sigma', C \vdash A_k,$$

where B and C are arbitrary formulas such that $\Sigma', B \vee C = \Sigma_k$.

The sequence 6) is called a *formal proof.* It is a formal proof of its last term $\Sigma_n \vdash A_n$.

We write $\Sigma \nvdash A$ for "not $\Sigma \vdash A$".

A scheme of formal deducibility may have various formal proofs. Perhaps, one may not know how to construct a formal proof of it. It is significant, however, that any proposed formal proof can be checked mechanically to decide whether it is indeed a formal proof of this scheme. This is done by checking whether the rules of formal deduction are correctly applied and whether the last term of the formal proof is identical with this scheme. In this sense, rules of formal deduction and formal proofs serve to clarify the concepts of rules of inference and proofs in informal reasoning.

Now the significance of "formal" has been explained in full. The word "formal" may sometimes be omitted if no confusion will arise.

Remarks

(1) Tautological consequence ($\Sigma \models A$) and formal deducibility ($\Sigma \vdash A$) are different matters. The former belongs to semantics, while the latter belongs to syntax.

(2) Both tautological consequence and formal deducibility are studied in the metalanguage by means of reasoning which is informal. (3) $\models$ and $\vdash$ are not symbols of $\mathcal{L}^p$. They should not be confused with $\rightarrow$, which is a symbol of $\mathcal{L}^p$, a connective used for forming formulas. But there is a connection between $\models$ (or $\vdash$) and $\rightarrow$ such that $A \models B$ iff $A \rightarrow B$ is a tautology and $A \vdash B$ iff $\emptyset \vdash A \rightarrow B$.

Definition 2.6.1 is an inductive one. We may compare this definition with Definition 2.2.2 of $Form(\mathcal{L}^p)$ to see that schemes of formal deducibility correspond to formulas, rules of formal deduction to formation rules, and formal proofs to formation sequences (see Exercise 2.2.1).

Statements concerning formal deducibility can be proved by induction on its structure (of generation). The basis of induction is to prove that

$$A \vdash A$$

which is generated directly by the rule (Ref), has a certain property. The induction step is to prove that the other ten rules preserve this property. For instance, in the case of ($\vee-$), we suppose

$$\Sigma, A \vdash C$$
$$\Sigma, B \vdash C$$

have this property (induction hypothesis) and want to prove that

$$\Sigma, A \vee B \vdash C$$

has also this property.

Theorem 2.6.2.
If $\Sigma \vdash A$, then there is some finite $\Sigma^\circ \subseteq \Sigma$ such that $\Sigma^\circ \vdash A$.

Proof. By induction on the structure of $\Sigma \vdash A$.
Basis. The premise A of $A \vdash A$ generated by (Ref) is itself finite.
Induction step. We distinguish ten cases.

Case of (+):

$$\begin{aligned} &\text{If} \quad \Sigma \vdash A, \\ &\text{then} \quad \Sigma,\ \Sigma' \vdash A. \end{aligned}$$

By the induction hypothesis, there is some finite $\Sigma^\circ \subseteq \Sigma$ such that $\Sigma^\circ \vdash A$. Σ° is also a finite subset of Σ, Σ'.

Case of ($\neg-$):

$$\begin{aligned} &\text{If} \quad \Sigma,\ \neg A \vdash B, \\ &\qquad\ \ \Sigma, \neg A \vdash \neg B, \\ &\text{then} \quad \Sigma \vdash A. \end{aligned}$$

First, we prove:

(1) There is some finite $\Sigma_1 \subseteq \Sigma$ such that $\Sigma_1,\ \neg A \vdash B$.

By the induction hypothesis, there is some finite $\Sigma' \subseteq \{\Sigma, \neg A\}$ such that $\Sigma' \vdash B$. By (+), we have Σ', $\neg A \vdash B$. Suppose $\neg A \notin \Sigma'$, then $\Sigma' \subseteq \Sigma$; we obtain (1) by setting $\Sigma_1 = \Sigma'$. Suppose $\neg A \in \Sigma'$, then $\Sigma' - \{\neg A\} \subseteq \Sigma$. We thus obtain (1) by setting $\Sigma_1 = \Sigma' - \{\neg A\}$.

Similarly, we can prove:

(2) There is some finite $\Sigma_2 \subseteq \Sigma$ such that $\Sigma_2,\ \neg A \vdash \neg B$.

By (+), we obtain from (1) and (2)

$$\begin{aligned} &\Sigma_1,\ \Sigma_2,\ \neg A \vdash B, \\ &\Sigma_1,\ \Sigma_2,\ \neg A \vdash \neg B. \end{aligned}$$

Then Σ_1, $\Sigma_2 \vdash A$, where Σ_1, Σ_2 is a finite subset of Σ.

Case of ($\rightarrow-$):

$$\begin{aligned} &\text{If} \quad \Sigma \vdash A \rightarrow B, \\ &\qquad\ \ \Sigma \vdash A, \\ &\text{then} \quad \Sigma \vdash B. \end{aligned}$$

By the induction hypothesis, there are finite subsets Σ_1 and Σ_2 of Σ such that $\Sigma_1 \vdash A \rightarrow B$ and $\Sigma_2 \vdash A$. By (+) we have

$$\begin{aligned} &\Sigma_1,\ \Sigma_2 \vdash A \rightarrow B, \\ &\Sigma_1,\ \Sigma_2 \vdash A. \end{aligned}$$

Then Σ_1, $\Sigma_2 \vdash B$, where Σ_1, Σ_2 is a finite subset of Σ.

The proof of the other cases is left to the reader. By the basis and induction step, this theorem is proved. □

In a scheme $\Sigma \vdash A$ of formal deducibility, the premise is a set of formulas, and the conclusion consists of one formula. When a number of schemes have the same premise, we stipulate to write $\Sigma \vdash \Sigma'$ for "for any $B \in \Sigma'$, $\Sigma \vdash B$". Thus when Σ' is infinite, $\Sigma \vdash \Sigma'$ consists of an infinite number of schemes.

Theorem 2.6.3.
[1] $\Sigma \vdash A$ in the case $A \in \Sigma$.
[2] If $\Sigma \vdash \Sigma'$,
$\Sigma' \vdash A$,
then $\Sigma \vdash A$. (*Transitivity* of *deducibility*)

Proof. [1] has been proved in an example. [2] is proved as follows:
(1) $\Sigma' \vdash A$ (by supposition).
(2) $A_1, \ldots, A_n \vdash A$, where $A_1, \ldots, A_n \in \Sigma'$ (by Thm 2.6.2, (1)).
(3) $A_1, \ldots, A_{n-1} \vdash A_n \to A$ (by $(\to +)$, (2)).
(4) $\emptyset \vdash A_1 \to (\ldots(A_n \to A)\ldots)$ (analogous to (3)).
(5) $\Sigma \vdash A_1 \to (\ldots(A_n \to A)\ldots)$ (by (+), (4)).
(6) $\Sigma \vdash A_1$ (by supposition and $A_1 \in \Sigma'$).
(7) $\Sigma \vdash A_2 \to (\ldots(A_n \to A)\ldots)$ (by $(\to -)$, (5), (6)).
(8) $\Sigma \vdash A_n \to A$ (analogous to (7)).
(9) $\Sigma \vdash A_n$ (by supposition and $A_n \in \Sigma'$).
(10) $\Sigma \vdash A$ (by $(\to -)$, (8), (9)). □

The rule of transitivity of deducibility is denoted by (Tr).

Remarks

Although $\Sigma \vdash \Sigma'$ in the supposition of (Tr) may contain an infinite number of schemes, we use in the proof only a finite number of them, since by Theorem 2.6.2, we have $A_1, \ldots, A_n \vdash A$ from $\Sigma' \vdash A$ ($A_1, \ldots, A_n \in \Sigma'$). Hence, when $\Sigma \vdash \Sigma'$ is written in a formal proof, it does not mean the formal proof contains infinite steps, which contradicts Definition 2.6.1.

Theorem 2.6.4.
[1] $A \to B, A \vdash B$.
[2] $A \vdash B \to A$.
[3] $A \to B, B \to C \vdash A \to C$.
[4] $A \to (B \to C), A \to B \vdash A \to C$.

Proof. [3] has been proved in an example. The proofs of [1], [2], and [4] are left to the reader. □

By definition, the terms in a formal proof should be generated by the rules of formal deduction. But in writing formal proofs we can use the demonstrated schemes of formal deducibility, because they can be reduced to the rules. Therefore, the rules are axioms of formal deduction, while the schemes are theorems.

Theorem 2.6.5.
[1] $\neg\neg \vdash$ A.
[2] If Σ, A $\vdash$ B,
Σ, A $\vdash \neg$B,
then $\Sigma \vdash \neg$A. (*Reductio ad absurdum*)
[3] A $\vdash \neg\neg$A.
[4] A, $\neg$A $\vdash$ B.
[5] A $\vdash \neg$A $\rightarrow$ B.
[6] $\neg$A $\vdash$ A $\rightarrow$ B.

Proof. We choose to prove [1], [2], and [3]. The rest are left to the reader.

Proof of [1].
(1) $\neg\neg$A, $\neg$A $\vdash \neg$A (by (ϵ)).
(2) $\neg\neg$A, $\neg$A $\vdash \neg\neg$A (by (ϵ)).
(3) $\neg\neg$A $\vdash$ A (by ($\neg-$), (1), (2)).

Proof of [2].
(1) Σ, A $\vdash$ B (by supposition).
(2) Σ, $\neg\neg$A $\vdash \Sigma$ (by (ϵ)).
(3) $\neg\neg$A $\vdash$ A (by Thm 2.6.5 [1]).
(4) Σ, $\neg\neg$A $\vdash$ A (by (+), (3)).
(5) Σ, $\neg\neg$A $\vdash$ B (by (Tr), (2), (4), (1)).
(6) Σ, $\neg\neg$A $\vdash \neg$B (analogous to (5)).
(7) $\Sigma \vdash \neg$A (by ($\neg-$), (5), (6)).

Proof of [3].
(1) A, $\neg$A $\vdash$ A (by (ϵ)).
(2) A, $\neg$A $\vdash \neg$A (by (ϵ)).
(3) A $\vdash \neg\neg$A (by Thm 2.6.5 [2], (1), (2)). □

The rule of *reductio ad absurdum* is denoted by $(\neg+)$. $(\neg+)$ and $(\neg-)$, the rule of indirect proof, are similar in shape, but different in strength. $(\neg-)$ is stronger than $(\neg+)$. $(\neg+)$ has been proved above. But, if $(\neg-)$ is replaced by $(\neg+)$ in the rules, then $(\neg-)$ cannot be proved. This concerns the notion of independence, which will be discussed in Chapter 5.

$(\neg+)$ is also called $\neg$*–introduction.*

Since $(\neg-)$ is stronger than $(\neg+)$, a scheme provable by $(\neg+)$ is necessarily provable by $(\neg-)$, but a scheme provable by $(\neg-)$ is not necessarily provable by $(\neg+)$. When $(\neg-)$ and $(\neg+)$ are both available, it is usually more convenient to use $(\neg+)$.

Theorem 2.6.6.
[1] $A \to B \vdash \neg B \to \neg A$.
[2] $A \to \neg B \vdash B \to \neg A$.
[3] $\neg A \to B \vdash \neg B \to A$.
[4] $\neg A \to \neg B \vdash B \to A$.
[5] If $A \vdash B$, then $\neg B \vdash \neg A$.
[6] If $A \vdash \neg B$, then $B \vdash \neg A$.
[7] If $\neg A \vdash B$, then $\neg B \vdash A$.
[8] If $\neg A \vdash \neg B$, then $B \vdash A$.

Proof. We choose to prove [1]:
(1) $A \to B, \neg B, A \vdash \neg B$ (by (ϵ)).
(2) $A \to B, A \vdash B$ (by Thm 2.6.4 [1]).
(3) $A \to B, \neg B, A \vdash B$ (by $(+)$, (2)).
(4) $A \to B, \neg B \vdash \neg A$ (by $(\neg+)$, (3), (1)).
(5) $A \to B \vdash \neg B \to \neg A$ (by $(\to+)$, (4)). □

Theorem 2.6.7.
[1] $\neg A \to A \vdash A$.
[2] $A \to \neg A \vdash \neg A$.
[3] $A \to B, A \to \neg B \vdash \neg A$.
[4] $A \to B, \neg A \to B \vdash B$.
[5] $\neg(A \to B) \vdash A$.
[6] $\neg(A \to B) \vdash \neg B$.

Proof. We choose to prove [1] and [6].
Proof of [1].

(1) $\neg A \to A,\ \neg A \vdash A$ (by Thm 2.6.4 [1]).
(2) $\neg A \to A,\ \neg A \vdash \neg A$ (by (ϵ)).
(3) $\neg A \to A \vdash A$ (by $(\neg -)$, (1), (2)).

Proof of [6].
(1) $\neg(A \to B), B \vdash \neg(A \to B)$ (by (ϵ)).
(2) $B \vdash A \to B$ (by Thm 2.6.4 [2]).
(3) $\neg(A \to B), B \vdash A \to B$ (by $(+)$, (2)).
(4) $\neg(A \to B) \vdash \neg B$ (by $(\neg +)$, (3), (1)). □

For two formulas A and B, we write

$$A \vdash\dashv B$$

for "$A \vdash B$ and $B \vdash A$". A and B are said to be *syntactically equivalent* (or simply *equivalent* if no confusion will arise) iff $A \vdash\dashv B$ holds.

We write $\dashv$ to denote the converse of $\vdash$.

Theorem 2.6.8.
[1] $A \wedge B \vdash A, B$.
[2] $A, B \vdash A \wedge B$.
[3] $A \wedge B \vdash\dashv B \wedge A$. ($\wedge$-*commutativity*)
[4] $(A \wedge B) \wedge C \vdash\dashv A \wedge (B \wedge C)$. ($\wedge$-*associativity*)
[5] $\neg(A \wedge B) \vdash\dashv A \to \neg B$.
[6] $\neg(A \to B) \vdash\dashv A \wedge \neg B$.
[7] $\emptyset \vdash \neg(A \wedge \neg A)$. (*Law of non-contradiction*)

Proof. We choose to prove [5].
Proof of $\vdash$ of [5].
(1) $A, B \vdash A \wedge B$ (by this theorem [2]).
(2) $\neg(A \wedge B), A, B \vdash A \wedge B$ (by $(+)$, (1)).
(3) $\neg(A \wedge B), A, B \vdash \neg(A \wedge B)$ (by (ϵ)).
(4) $\neg(A \wedge B), A \vdash \neg B$ (by $(\neg +)$, (2), (3)).
(5) $\neg(A \wedge B) \vdash A \to \neg B$ (by $(\to +)$, (4)).

Proof of $\dashv$ of [5].
(1) $A \wedge B \vdash A$ (by this theorem [1]).
(2) $A \to \neg B, A \wedge B \vdash A$ (by $(+)$, (1)).
(3) $A \to \neg B, A \wedge B \vdash B$ (analogous to (2)).

(4) $A \to \neg B, A \wedge B \vdash A \to \neg B$ (by (ϵ)).
(5) $A \to \neg B, A \wedge B \vdash \neg B$ (by ($\to -$), (4), (2)).
(6) $A \to \neg B \vdash \neg(A \wedge B)$ (by ($\neg +$), (3), (5)). □

Theorem 2.6.9.
[1] $A \vdash A \vee B, B \vee A$.
[2] $A \vee B \dashv\vdash B \vee A$. ($\vee$*-commutativity*)
[3] $(A \vee B) \vee C \dashv\vdash A \vee (B \vee C)$. ($\vee$*-associativity*)
[4] $A \vee B \dashv\vdash \neg A \to B$.
[5] $A \to B \dashv\vdash \neg A \vee B$.
[6] $\neg(A \vee B) \dashv\vdash \neg A \wedge \neg B$. (*De Morgan's Law*)
[7] $\neg(A \wedge B) \dashv\vdash \neg A \vee \neg B$. (*De Morgan's Law*)
[8] $\emptyset \vdash A \vee \neg A$. (*Law of excluded middle*)

Proof. We choose to prove [2] and [4].
Proof of $\vdash$ of [2]. ($\dashv$ of [2] is the same as $\vdash$.)
(1) $A \vdash B \vee A$ (by this theorem [1]).
(2) $B \vdash B \vee A$ (by this theorem [1]).
(3) $A \vee B \vdash B \vee A$ (by ($\vee -$), (1), (2)).

Proof of $\vdash$ of [4].
(1) $A \vdash \neg A \to B$ (by Thm 2.6.5 [5]).
(2) $B \vdash \neg A \to B$ (by Thm 2.6.4 [2]).
(3) $A \vee B \vdash \neg A \to B$ (by ($\vee -$), (1), (2)).

Proof of $\dashv$ of [4].
(1) $\neg A \to B, \neg(A \vee B), A \vdash A$ (by (ϵ)).
(2) $\neg A \to B, \neg(A \vee B), A \vdash A \vee B$ (by ($\vee +$), (1)).
(3) $\neg A \to B, \neg(A \vee B), A \vdash \neg(A \vee B)$ (by (ϵ)).
(4) $\neg A \to B, \neg(A \vee B) \vdash \neg A$ (by ($\neg +$), (2), (3)).
(5) $\neg A \to B, \neg(A \vee B) \vdash \neg A \to B$ (by (ϵ)).
(6) $\neg A \to B, \neg(A \vee B) \vdash B$ (by ($\to -$), (5), (4)).
(7) $\neg A \to B, \neg(A \vee B) \vdash A \vee B$ (by ($\vee +$), (6)).
(8) $\neg A \to B, \neg(A \vee B) \vdash \neg(A \vee B)$ (by (ϵ)).
(9) $\neg A \to B \vdash A \vee B$ (by ($\neg -$), (7), (8)). □

Note that in the above proof, (2) is distinct from (7), and (3) is distinct from (8).

Up to now the steps of formal proofs have been written in detail. Henceforth, some of them will be omitted for simplicity since they are more or less obvious. The justifications for formal proofs will be omitted as well.

Theorem 2.6.10.

[1] $A \vee (B \wedge C) \vdash\dashv (A \vee B) \wedge (A \vee C)$.
[2] $A \wedge (B \vee C) \vdash\dashv (A \wedge B) \vee (A \wedge C)$.
[3] $A \to (B \wedge C) \vdash\dashv (A \to B) \wedge (A \to C)$.
[4] $A \to (B \vee C) \vdash\dashv (A \to B) \vee (A \to C)$.
[5] $(A \wedge B) \to C \vdash\dashv (A \to C) \vee (B \to C)$.
[6] $(A \vee B) \to C \vdash\dashv (A \to C) \wedge (B \to C)$.

The proof of Theorem 2.6.10 is left to the reader.

Theorem 2.6.11.

[1] $A \leftrightarrow B, A \vdash B$.
$A \leftrightarrow B, B \vdash A$.
[2] $A \leftrightarrow B \vdash\dashv B \leftrightarrow A$. ($\leftrightarrow$-*commutativity*)
[3] $A \leftrightarrow B \vdash\dashv \neg A \leftrightarrow \neg B$.
[4] $\neg(A \leftrightarrow B) \vdash\dashv A \leftrightarrow \neg B$.
[5] $\neg(A \leftrightarrow B) \vdash\dashv \neg A \leftrightarrow B$.
[6] $A \leftrightarrow B \vdash\dashv (\neg A \vee B) \wedge (A \vee \neg B)$.
[7] $A \leftrightarrow B \vdash\dashv (A \wedge B) \vee (\neg A \wedge \neg B)$.
[8] $(A \leftrightarrow B) \leftrightarrow C \vdash\dashv A \leftrightarrow (B \leftrightarrow C)$ ($\leftrightarrow$-*associativity*)
[9] $A \leftrightarrow B, B \leftrightarrow C \vdash A \leftrightarrow C$.
[10] $A \leftrightarrow \neg A \vdash B$.
[11] $\emptyset \vdash (A \leftrightarrow B) \vee (A \leftrightarrow \neg B)$.

The proof of Theorem 2.6.11 is left to the reader.

$A \leftrightarrow B$ may be considered as $(A \to B) \wedge (B \to A)$. Therefore the rules of deduction concerning $\leftrightarrow$ may be stated as follows:

$$
\begin{array}{ll}
(\leftrightarrow -) & \text{If } \Sigma \vdash A \leftrightarrow B, \\
& \text{then } \Sigma \vdash A \to B, \\
& \qquad \Sigma \vdash B \to A. \\
(\leftrightarrow +) & \text{If } \Sigma \vdash A \to B, \\
& \qquad \Sigma \vdash B \to A, \\
& \text{then } \Sigma \vdash A \leftrightarrow B.
\end{array}
$$

The following lemma and three theorems correspond respectively to Lemma 2.5.3 and Theorems 2.5.4, 2.5.5, and 2.5.2 in the last section.

Lemma 2.6.12.
If $A \vdash\dashv A'$ and $B \vdash\dashv B'$, then
[1] $\neg A \vdash\dashv \neg A'$.
[2] $A \wedge B \vdash\dashv A' \wedge B'$.
[3] $A \vee B \vdash\dashv A' \vee B'$.
[4] $A \rightarrow B \vdash\dashv A' \rightarrow B'$.
[5] $A \leftrightarrow B \vdash\dashv A' \leftrightarrow B'$. □

Theorem 2.6.13. (***Replaceability of equivalent formulas***)
If $B \vdash\dashv C$ and A' result from A by replacing some (not necessarily all) occurrences of B in A by C, then $A \vdash\dashv A'$. □

Theorem 2.6.14. (***Duality***)
Suppose A is a formula composed of atoms and the connectives $\neg, \wedge$, and $\vee$ by the formation rules concerned and A' is the dual of A. Then $A' \vdash\dashv \neg A$. □

Theorem 2.6.15.
[1] $A_1, \ldots, A_n \vdash A$ iff $\emptyset \vdash A_1 \wedge \ldots \wedge A_n \rightarrow A$.
[2] $A_1, \ldots, A_n \vdash A$ iff $\emptyset \vdash A_1 \rightarrow (\ldots(A_n \rightarrow A)\ldots)$. □

When the premise is empty, we have the special case $\emptyset \vdash A$ of formal deducibility. Obviously $\emptyset \vdash A$ iff $\Sigma \vdash A$ for any Σ.

It has been mentioned before that A is said to be formally provable from Σ when $\Sigma \vdash A$ holds. Now A is said to be *formally provable* when $\emptyset \vdash A$ holds. The laws of non-contradiction $\neg(A \wedge \neg A)$ and excluded middle $A \vee \neg A$ are instances of formally provable formulas.

By Theorem 2.6.2, the premise Σ of $\Sigma \vdash A$ can be reduced to a finite set; and by Theorem 2.6.15, $A_1, \ldots, A_n \vdash A$ is equivalent to a formally provable formula. Hence the formal deducibility between Σ and A can be expressed, in a sense, by a formally provable formula. The significance of formally provable formulas will be seen in the discussion of soundness and completeness in Chapter 5.

Since the rules of formal deduction (for propositional logic in this chapter, and similarly for first-order logic, constructive logic, and modal logic in later chapters) express naturally and intuitively the rules of informal reasoning, the formal deduction based upon these rules is called *natural deduction*. There is another type of formal deduction, which will be introduced in Chapter 4.

It has been seen that to write formal proofs out in full is rather tedious because the same formulas are often used repeatedly. A simpler and clearer form of formal proofs to facilitate writing and reading will be introduced in the Appendix.

Exercises 2.6.

2.6.1. Prove Theorem 2.6.9 [5], [6].

2.6.2. Prove Theorem 2.6.11 [4], [8], [10], [11].

2.6.3. Prove the following:

[1] $(A \to B) \to B \vdash (B \to A) \to A$.

[2] $(A \to B) \to C \vdash (A \to C) \to C$.

[3] $(A \to B) \to C \vdash (C \to A) \to (A_1 \to A)$.

[4] $A \land \neg B \to A_1 \lor C, B \to \neg A, A \to \neg C \vdash A \to A_1$.

2.6.4. Prove $(\neg-)$ by $(\neg+)$ and the following:

[1] If $\Sigma \vdash \neg\neg A$, then $\Sigma \vdash A$.

2.6.5. Prove $(\neg-)$ by (Ref), $(+)$, $(\to+)$, and the following:

[1] If $\Sigma \vdash \neg\neg A$, then $\Sigma \vdash A$.

[2] If $\Sigma \vdash A$, then $\Sigma \vdash \neg\neg A$.

[3] If $\Sigma \vdash A \to B, \neg B$, then $\Sigma \vdash \neg A$.

2.6.6. Prove $(\neg-)$ by (Ref), $(+)$, $(\to+)$, and the following:

[1] If $\Sigma \vdash \neg\neg A$, then $\Sigma \vdash A$.

[2] If $\Sigma \vdash A \to \neg B, B$, then $\Sigma \vdash \neg A$.

2.6.7. Prove $(\neg-)$ by (Ref), $(+)$, $(\to+)$, and the following:

[1] If $\Sigma \vdash A$, then $\Sigma \vdash \neg\neg A$.

[2] If $\Sigma \vdash \neg A \to B, \neg B$, then $\Sigma \vdash A$.

2.6.8. Prove $(\neg-)$ by (Ref), $(+)$, $(\to+)$, and the following:

[1] If $\Sigma \vdash \neg A \to \neg B, B$, then $\Sigma \vdash A$.

2.7. DISJUNCTIVE AND CONJUNCTIVE NORMAL FORMS

Formulas can be transformed into normal forms so that they become more convenient for symbol manipulations. In this section, two kinds of normal forms in propositional logic will be discussed: the disjunctive and conjunctive normal forms.

Definition 2.7.1. (***Literal, clause***)
Atoms and their negations are called *literals.*
Disjunctions (conjunctions) with literals as disjuncts (conjuncts) are called *disjunctive (conjunctive) clauses.* Disjunctive and conjunctive clauses are simply called *clauses.*

Definition 2.7.2. (***Disjunctive, conjunctive normal form***)
A disjunction with conjunctive clauses as its disjuncts is called a *disjunctive normal form.*
A conjunction with disjunctive clauses as its conjuncts is called a *conjunctive normal form.*

Disjunctive and conjunctive normal forms are respectively of the following forms:

$$(\mathrm{A}_{11} \wedge \ldots \wedge \mathrm{A}_{1n_1}) \vee \ldots \vee (\mathrm{A}_{k1} \wedge \ldots \wedge \mathrm{A}_{kn_k})$$

$$(\mathrm{A}_{11} \vee \ldots \vee \mathrm{A}_{1n_1}) \wedge \ldots \wedge (\mathrm{A}_{k1} \vee \ldots \vee \mathrm{A}_{kn_k})$$

where A_{ij} ($i = 1, \ldots, k$; $j = 1, \ldots, n_i$) are literals.

Example
Observe the following formulas:
(1) p
(2) $\neg\mathrm{p} \vee \mathrm{q}$
(3) $\neg\mathrm{p} \wedge \mathrm{q} \wedge \neg\mathrm{r}$
(4) $\neg\mathrm{p} \vee (\mathrm{q} \wedge \neg\mathrm{r})$
(5) $\neg\mathrm{p} \wedge (\mathrm{q} \vee \neg\mathrm{r}) \wedge (\neg\mathrm{q} \vee \mathrm{r})$

(1) is an atom, and therefore a literal. It is a disjunction with only one disjunct. It is also a conjunction with only one conjunct. Hence it is a disjunctive or conjunctive clause with one literal. It is a disjunctive normal

form with one conjunctive clause p. It is also a conjunctive normal form with one disjunctive clause p.

(2) is a disjunction with two disjuncts, and a disjunctive normal form with two clauses, each with one literal. It is also a conjunction with one conjunct, and a conjunctive normal form with one clause which consists of two literals.

Similarly, (3) is a conjunction and a conjunctive normal form. It is also a disjunction and a disjunctive normal form.

(4) is a disjunctive normal form, but not a conjunctive one.

(5) is a conjunctive normal form, but not a disjunctive one.

If $\vee$ is exchanged for $\wedge$ in (4) and (5), then (4) becomes a conjunctive normal form and (5) a disjunctive one.

Theorem 2.7.3. Any A $\in$ *Form* $(\mathcal{L}^p)$ is tautologically equivalent to some disjunctive normal form.

Proof. If A is a contradiction, A is tautologically equivalent to the disjunctive normal form $p \wedge \neg p$, p being any atom occurring in A.

If A is not a contradiction, we can without loss of generality prove the theorem by considering an instance of A. Suppose A is a formula with three atoms p, q, and r occurring in it, and the value of A is 1 iff 1, 1, 0, or 1, 0, 1 or 0, 0, 1 are assigned to p, q, r respectively. (A has the value 1 for at least one such assignment.)

For each of the above assignments, we form a conjunctive clause with three literals, each being one of the atoms or its negation according to whether this atom is assigned 1 or 0. Therefore we form in order, for the above assignments, the following three conjunctive clauses:

$$p \wedge q \wedge \neg r, \tag{1}$$

$$p \wedge \neg q \wedge r, \tag{2}$$

$$\neg p \wedge \neg q \wedge r. \tag{3}$$

Obviously (1) has the value 1 iff 1, 1, 0 are assigned to p, q, r; (2) has the value 1 iff 1, 0, 1 are assigned; (3) has the value 1 iff 0, 0, 1 are assigned. Therefore the following disjunctive normal form (with (1), (2), and (3) as clauses):

$$(p \wedge q \wedge \neg r) \vee (p \wedge \neg q \wedge r) \vee (\neg p \wedge \neg q \wedge r)$$

is tautologically equivalent to A.

For a tautology A, the required disjunctive normal form may simply be $p \vee \neg p$, where p is any atom occurring in A. □

Theorem 2.7.4.

Any $A \in Form(\mathcal{L}^p)$ is tautologically equivalent to some conjunctive normal form.

Proof. Analogous to that of Theorem 2.7.3, with modifications. □

Remarks

After reading the Completeness Theorem stated in Chapter 5, we can verify that the disjunctive and conjunctive normal forms are also syntactically equivalent to the original formula.

A disjunctive (conjunctive) normal form equivalent to a formula A is called a disjunctive (conjunctive) normal form of A.

A formula and its negation are called *complementary* formulas, each being the *complement* of the other.

The following theorem and corollary obviously hold.

Theorem 2.7.5.

A disjunctive normal form is a contradiction, iff complementary literals occur in each of its (conjunctive) clauses.

A conjunctive normal form is a tautology, iff complementary literals occur in each of its (disjunctive) clauses. □

Corollary 2.7.6.

A formula is a contradiction, iff complementary literals occur in each of the (conjunctive) clauses of its disjunctive normal form.

A formula is a tautology, iff complementary literals occur in each of the (disjunctive) clauses of its conjunctive normal form. □

A *full disjunctive* or *conjunctive normal form* of a formula A is one which contains all the atoms of A in each of its clauses, every atom occurring only once in each clause (in the form of an atom or its negation), and the clauses of which are all different.

If A is neither a tautology nor a contradiction, the normal forms of A formed in the proofs of Theorems 2.7.3 and 2.7.4 are full disjunctive and conjunctive normal forms.

We now introduce another method of forming normal forms of formulas. We have the following tautological equivalences which can easily be proved:

1) $A \to B \models\!\dashv \neg A \vee B$.
2) $A \leftrightarrow B \models\!\dashv (\neg A \vee B) \wedge (A \vee \neg B)$.
3) $A \leftrightarrow B \models\!\dashv (A \wedge B) \vee (\neg A \wedge \neg B)$.
4) $\neg\neg A \models\!\dashv A$.
5) $\neg(A_1 \wedge \ldots \wedge A_n) \models\!\dashv \neg A_1 \vee \ldots \vee \neg A_n$.
6) $\neg(A_1 \vee \ldots \vee A_n) \models\!\dashv \neg A_1 \wedge \ldots \wedge \neg A_n$.
7) $A \wedge (B_1 \vee \ldots \vee B_n) \models\!\dashv (A \wedge B_1) \vee \ldots \vee (A \wedge B_n)$.
   $(B_1 \vee \ldots \vee B_n) \wedge A \models\!\dashv (B_1 \wedge A) \vee \ldots \vee (B_n \wedge A)$.
8) $A \vee (B_1 \wedge \ldots \wedge B_n) \models\!\dashv (A \vee B_1) \wedge \ldots \wedge (A \vee B_n)$.
   $(B_1 \wedge \ldots \wedge B_n) \vee A \models\!\dashv (B_1 \vee A) \wedge \ldots \wedge (B_n \vee A)$.

By the replaceability of tautological equivalences, we can replace the above formulas on the left with the corresponding ones on the right to yield a formula tautologically equivalent to the original one. By 1)–3) we eliminate $\to$ and $\leftrightarrow$. By 4)–6) we eliminate $\neg$, $\wedge$, and $\vee$ from the scope of $\neg$, such that any $\neg$ has an atom as its scope. By 7) we eliminate $\vee$ from the scope of $\wedge$, and by 8), $\wedge$ from the scope of $\vee$. Disjunctive and conjunctive normal forms are then obtained.

Certain tautological equivalences may be used to simplify the transformation process or to obtain simpler normal forms. For instance, the following tautological equivalences

$$A \vee A \models\!\dashv A$$

$$A \wedge A \models\!\dashv A$$

can be used to delete the redundant disjuncts and conjuncts. Such redundant disjuncts and conjuncts may be literals in clauses, or clauses in normal forms. By

$$A \vee (A \wedge B) \models\!\dashv A,$$

$$A \wedge (A \vee B) \models\!\dashv A,$$

if all the literals in one clause of a normal form occur in another clause, the longer clause can be deleted. By

$$A \vee (B \wedge \neg B \wedge C) \models\!\dashv A,$$

$$A \wedge (B \vee \neg B \vee C) \models\!\dashv A,$$

the clauses with complementary literals can be deleted in normal forms.

We can easily verify that syntactically equivalent formulas are obtained by replacing $\models\!\dashv$ with $\vdash\!\dashv$ in the above tautological equivalences. Therefore, by the replaceability of syntactical equivalences, the normal forms obtained by the method described above are also syntactically equivalent to the original formulas.

The problem of simplification of normal forms will not be discussed in this book.

Exercises 2.7.

2.7.1. Prove Theorem 2.7.4.

2.7.2. Find the disjunctive and conjunctive normal forms of the following formulas:

[1] $(A \to A \lor B) \to B \land C \leftrightarrow \neg A \land C$

[2] $(A \leftrightarrow B \land A \lor \neg C) \to (A \land \neg B \to C)$

[3] $(A \leftrightarrow B) \leftrightarrow [(\neg A \leftrightarrow C) \to (B \leftrightarrow \neg C)]$

[4] $\neg(A \land \neg A)$

2.7.3. Suppose A is a non-contradiction with n distinct atoms in it, and B is a full disjunctive normal form of A. Prove that A is a tautology iff the number of clauses in B equals 2^n.

2.8. ADEQUATE SETS OF CONNECTIVES

Formulas $A \to B$ and $\neg A \lor B$ are tautologically equivalent. Then $\to$ is said to be *definable* in terms of (or *reducible* to) $\neg$ and $\lor$. Similarly $\lor$ is definable in terms of $\neg$ and $\to$, because $A \lor B$ is tautologically equivalent to $\neg A \to B$.

We have up to now mentioned one unary and four binary connectives. In fact there are more unary and binary connectives, and also n-ary connectives for $n > 2$.

In this section we shall use two italic small Latin letters f and g (with or without subscripts) to denote any connectives. We shall write

$$fA_1 \ldots A_n$$

for the formula formed by an n-ary connective f connecting formulas A_1, $\ldots$, A_n.

Two n-ary ($n \geq 1$) connectives are identical iff they have the same truth tables. Hence for any $n \geq 1$, there are $2^{(2^n)}$ distinct n-ary connectives. For instance, there are $2^{(2^1)} = 4$ distinct unary and $2^{(2^2)} = 16$ distinct binary connectives.

Suppose f_1, f_2, f_3, and f_4 are distinct unary connectives. They have the following truth tables:

A	f_1A	f_2A	f_3A	f_4A
1	1	1	0	0
0	1	0	1	0

where f_3 is negation.

Suppose g_1, $\ldots$, g_{16} are distinct binary connectives. Their truth tables are as follows:

A	B	g_1AB	g_2AB	g_3AB	g_4AB	g_5AB	g_6AB	g_7AB	g_8AB
1	1	1	1	1	1	0	1	1	1
1	0	1	1	1	0	1	1	0	0
0	1	1	1	0	1	1	0	1	0
0	0	1	0	1	1	1	0	0	1

g_9AB	g_{10}AB	g_{11}AB	g_{12}AB	g_{13}AB	g_{14}AB	g_{15}AB	g_{16}AB
0	0	0	1	0	0	0	0
1	1	0	0	1	0	0	0
1	0	1	0	0	1	0	0
0	1	1	0	0	0	1	0

where g_2, g_4, g_8, and g_{12} are $\vee$, $\rightarrow$, $\leftrightarrow$, and $\wedge$ respectively.

g_5 is called *Sheffer stroke*, usually denoted by $|$; g_{15} is usually denoted by $\downarrow$.

One of the $2^{(2^3)} = 256$ ternary connectives is *if-then-else* which has the following truth table:

A	B	C	If A then B else C.
1	1	1	1
1	1	0	1
1	0	1	0
1	0	0	0
0	1	1	1
0	1	0	0
0	0	1	1
0	0	0	0

A set of connectives is said to be *adequate* iff any n-ary ($n \geq 1$) connective can be defined in terms of them.

Suppose f is any n-ary connective. By the method used in the proof of Theorem 2.7.3, we can obtain a disjunctive normal form tautologically equivalent to $f\mathrm{p}_1 \ldots \mathrm{p}_n$, which is formed by f connecting the atoms p_1, $\ldots$, p_n. Note that only $\neg$, $\wedge$, and $\vee$ occur in a disjunctive form. Hence we have the following theorem.

Theorem 2.8.1.
$\{\neg, \wedge, \vee\}$ is an adequate set of connectives. □

Corollary 2.8.2.
$\{\neg, \vee\}$, $\{\neg, \wedge\}$, and $\{\neg, \to\}$ are adequate. □

Now we turn to consider propositional logic based not upon the five common connectives, but upon an adequate set of connectives, for instance, $\{\neg, \wedge\}$.

Let $\mathcal{L}^p_\circ$ be a sublanguage of $\mathcal{L}^p$, obtained by deleting from $\mathcal{L}^p$ three connectives $\vee$, $\to$, and $\leftrightarrow$. $Form(\mathcal{L}^p_\circ)$ is the set of formulas of $\mathcal{L}^p_\circ$. Obviously $Form(\mathcal{L}^p_\circ) \subseteq Form(\mathcal{L}^p)$.

Suppose $\Sigma \subseteq Form(\mathcal{L}^p_\circ)$ and $\mathrm{A} \in Form(\mathcal{L}^p_\circ)$, and $\vdash_\circ$ is the formal deducibility defined by the rules (Ref), (+), ($\neg-$), ($\wedge-$), and ($\wedge+$). Then $\Sigma \vdash_\circ \mathrm{A}$ implies $\Sigma \vdash \mathrm{A}$.

For $\mathrm{A} \in Form(\mathcal{L}^p)$ and $\Sigma \subseteq Form(\mathcal{L}^p)$, we define (by recursion) their translation $\mathrm{A}_\circ$ and $\Sigma_\circ$ into $\mathcal{L}^p_\circ$ as follows:

$$\begin{aligned}
&A_\circ = A \text{ for atom A},\\
&(\neg A)_\circ = \neg A_\circ,\\
&(A \wedge B)_\circ = A_\circ \wedge B_\circ,\\
&(A \vee B)_\circ = \neg(\neg A_\circ \wedge \neg B_\circ),\\
&(A \to B)_\circ = \neg(A_\circ \wedge \neg B_\circ),\\
&(A \leftrightarrow B)_\circ = (A \to B)_\circ \wedge (B \to A)_\circ\\
&\qquad\qquad = \neg(A_\circ \wedge \neg B_\circ) \wedge \neg(\neg A_\circ \wedge B_\circ),\\
&\Sigma_\circ = \{A_\circ \mid A \in \Sigma\}.
\end{aligned}$$

Then we have the following

Theorem 2.8.3.
Suppose $\Sigma \subseteq Form(\mathcal{L}^p)$ and $A \in Form(\mathcal{L}^p)$. Then $\Sigma \vdash A$ iff $\Sigma_\circ \vdash_\circ A_\circ$.

The proof of Theorem 2.8.3 is left to the reader.

Exercises 2.8.

2.8.1. The following sets of connectives are adequate:
[1] $\{\to, g_{14}\}$
[2] $\{\mid\}$
[3] $\{\downarrow\}$

2.8.2. $\{\wedge, \vee\}$ is not adequate.

2.8.3. $\{\leftrightarrow, g_9\}$ is not adequate.

2.8.4. $\wedge$ and $\vee$ cannot be defined in terms of $\neg$.

2.8.5. $\leftrightarrow$ cannot be defined in terms of $\to$.

2.8.6. $\mid$ and $\downarrow$ are the only binary connectives which are adequate by themselves.

2.8.7. Prove Theorem 2.8.3.

3

CLASSICAL FIRST-ORDER LOGIC

In propositional logic only the logical forms of compound propositions are analysed. A simple proposition is an unanalysed whole which is either true or false. For instance, in the following inference:

$$\begin{cases} \text{For any natural number } n \text{ there is a prime number} \\ \qquad \text{greater than } n. \text{ (Premise)} \\ 2^{100} \text{ is a natural number. (Premise)} \\ \text{There is a prime number greater than } 2^{100}. \text{ (Conclusion)} \end{cases}$$

the premises and conclusion are simple propositions. The correctness of this inference depends upon the interrelations of their logical forms. In propositional logic, however, the logical forms of simple propositions are not analysed and hence the correctness of this inference cannot be explained.

In this chapter classical first-order logic will be constructed, in which the logical forms of simple propositions will be analysed. In first-order logic, connectives and quantifiers will be used to form more complicated propositions, and the deducibility relations about them will be studied.

First-order logic is known by various names: predicate logic, elementary logic, the restricted predicate calculus, the restricted functional calculus, relational calculus, the theory of quantification with equality, etc. First-order logic seems to be in fashionable usage today.

3.1. PROPOSITION FUNCTIONS AND QUANTIFIERS

When we develop a scientific theory, there are typically a domain of objects, called individuals, which we intend to study. The domain is a non-empty set. Relations and functions on the domain are also studied.

Certain individuals, relations, and functions may be designated in terms of which others are defined. Then the four ingredients (a domain, the designated individuals, relations, and functions) constitute a structure. For instance, the structure $\mathcal{N}$ of natural numbers consists of the domain N (the set of natural numbers), the designated individual 0 (zero), the relation $=$ (equality), and the functions $'$ (successor), $+$ (addition), and $\cdot$ (multiplication), written as

$$\mathcal{N} = \langle N, 0, =, ', +, \cdot \rangle.$$

The structure $\mathcal{G}$ of a group is

$$\mathcal{G} = \langle G, e, =, \cdot \rangle$$

where G is the domain, which is a non-empty set (its members being the elements of the group), e is the designated individual which is the unit element, $=$ is the equality relation, and $\cdot$ is the operation of multiplication.

Various structures are studied in mathematics. Mathematical propositions are concerned with the domain, designated individuals, relations, and functions of the structure studied.

Variables ranging over the domain are used, for instance, in making general statements about the individuals such as

For all x, $x^2 \geq 0$.
For all x and y, $x^2 - y^2 = (x + y)(x - y)$.

and in expressing conditions which the individuals may or may not satisfy. For instance,

$$x + x = x \cdot x$$

expresses the condition that when the value of x is added to itself the result is the same as when it is multiplied by itself. This condition is satisfied only by 0 and 2. Similarly,

$$x \cdot x = y$$

is satisfied iff the value assigned to y is the square of that assigned to x.

Connectives are still used in forming compound propositions.

In addition, the terms "for all" (or "for every") and "there exists some" (or " there is some") are frequently used in mathematical propositions. They occur, for instance, in the definition of a limit $\lim_{x \to a} f(x) = b$:

For every $\varepsilon > 0$, there exists some $\delta > 0$ such that
if $|x - a| < \delta$ then $|f(x) - b| < \varepsilon$.

These terms are quantifiers. "For all", the *universal quantifier*, signifies the whole of the domain; "there exists some", the *existential quantifier*, signifies a part of it, meaning that there exists (at least) one individual in the domain having a certain property.

The significance of the quantifiers calls for some explanations. Suppose N is the domain. Then

4 is even.

5 is even.

are propositions, 4 and 5 being individuals in N. Replacing 4 and 5 by a variable x ranging over N, we obtain

1) x is even.

which is not a proposition and has no truth value, because x is a variable. 1) is a kind of function which is defined on the domain and becomes a proposition when some individual is assigned to x as its value. 1) is called a proposition function.

An n-ary *proposition function* on a domain D is an n-ary function mapping D^n into $\{1, 0\}$. 1) is a unary proposition function on N.

Prefixing a quantifier of x to 1), we obtain

2) For every x, x is even.

3) There is some x such that x is even.

Since x ranges over N, 2) and 3) mean, respectively,

4) For every natural number x, x is even.

5) There is some natural number x such that x is even.

The above are propositions.

Obviously the meaning of the variable x in 2) and 3) differs from that in 1). The x in 2) and 3) is no longer a variable ranging over the domain.

It has been quantified. Variables occurring in proposition functions are *free variables*, and quantified variables are *bound variables*. Free variables are real, while bound variables are apparent.

Another example of quantification: suppose N is still the domain, then

6) x divides y.

is a binary proposition function on N. Quantifying y universally in 6), we obtain

7) For every y, x divides y.

in which x is still free. 7) means

x divides every natural number.

7) is a unary proposition function, the value of which is determined only by x. Obviously 7) becomes a true proposition iff 1 is assigned to x as its value. By quantifying x universally in 7), we obtain the false proposition

For every x and y, x divides y.

The universal and existential quantifiers may be interpreted respectively as generalization of conjunction and disjunction. In case the domain D is finite, say $D = \{\alpha_1, \ldots, \alpha_k\}$, the following equivalences hold:

For every x $R(x)$ iff
$R(\alpha_1)$ and $\ldots$ and $R(\alpha_k)$.
There is some x such that $R(x)$ iff
$R(\alpha_1)$ or $\ldots$ or $R(\alpha_k)$.

where R is a property. However, if we want to make such statements about an infinite domain, it is natural to use quantifiers for doing this.

The range of quantifiers may be restricted to a subset of the domain. Suppose the domain is the set of real numbers. Consider the following statements:

8) For every $x \neq 0, x^2 > 0$.

9) There is some $x < 0$ such that $x^2 < 0$.

Note that the range of the universal quantifier in 8) is restricted by "$\neq 0$" to the subset of non-zero real numbers, and that of the existential quantifier in 9) is restricted by "< 0" to the subset of negative real numbers. They are called *restricted quantifiers* or quantifiers with restricted ranges. Statements with restricted quantifiers assert that every or some individual in a certain category of the domain has a certain property.

8) and 9) can be restated as follows:

10) For every x, if $x \neq 0$ then $x^2 > 0$.

11) There is some x such that $x < 0$ and $x^2 < 0$.

8) means

$$\{x \mid x \neq 0\} \subseteq \{x \mid x^2 > 0\}.$$

which is equivalent to 10). 9) means

There is some x such that $x < 0$ and at the same time $x^2 < 0$.

which is equivalent to 11).

Note the two patterns 10) and 11) for translating the restricted quantifiers. In 10) the universal quantifier is used with implication, while in 11) the existential quantifier is used with conjunction.

10) must not be replaced by

For every x, $x \neq 0$ and $x^2 > 0$.

which is obviously stronger than 8). 11) must not be replaced by

There is some x such that if $x < 0$ then $x^2 < 0$.

which is weaker than 9). 9) is false because there is no real number $x < 0$ such that $x^2 < 0$, but any non-negative real number makes "if $x < 0$ then $x^2 < 0$" vacuously true.

In first-order logic, the variables range over individuals of the domain. The quantifiers are interpreted in the familiar way as "for all individuals of the domain" and "there exists some individual of the domain".

In second-order logic, variables for subsets of the domain and its Cartesian products (that is, variables for relations on the domain) are allowed.

For instance, in the following propositions,

> Each non-empty set of natural numbers
> has a smallest element.
> Each bounded non-empty set of real numbers
> has a supremum.

we have to take all subsets of the domain into consideration and require variables and quantifiers for sets. In higher-order logic, variables and quantifiers for sets of sets, sets of sets of sets, etc. will be allowed.

3.2. FIRST-ORDER LANGUAGE

First-order language $\mathcal{L}$ is the formal language for first-order logic.

$\mathcal{L}$ may or may not be associated with a structure. First-order language (in a general sense), which is not associated with any structure, consists of eight classes of symbols.

Individual symbols include an infinite sequence of symbols. We use the roman-type small Latin letters

$$\mathrm{a} \quad \mathrm{b} \quad \mathrm{c}$$

(with or without subscripts or superscripts) for any individual symbol.

Relation symbols include an infinite sequence of symbols. The roman-type capital Latin letters

$$\mathrm{F} \quad \mathrm{G} \quad \mathrm{H}$$

(with or without subscripts or superscripts) are used for any relation symbol. Relation symbols are classified as unary, binary, etc.

There is a special binary relation symbol called the *equality symbol*, written as $\approx$. $\mathcal{L}$ may or may not contain $\approx$. $\mathcal{L}$ containing $\approx$ is called the first-order language with equality. In order to emphasize the speciality of the equality symbol, we stipulate that it is denoted by none of F, G, and H.

Function symbols are an infinite sequence of symbols. The roman-type small Latin letters

$$\mathrm{f} \quad \mathrm{g} \quad \mathrm{h}$$

(with or without subscripts or superscripts) are used for any function symbol. Function symbols are also classified as unary, binary, etc.

Free variable symbols and *bound variable symbols* are two infinite sequences of symbols. We use the roman-type small Latin letters (with or without subscripts or superscripts)

$$\mathrm{u} \quad \mathrm{v} \quad \mathrm{w}$$

for any free variable symbol, and

$$\mathrm{x} \quad \mathrm{y} \quad \mathrm{z}$$

for any bound variable symbol.

Connectives

$$\neg \quad \wedge \quad \vee \quad \rightarrow \quad \leftrightarrow$$

are the same as in Chapter 2.

The quantifier symbols

$$\forall \quad \exists$$

are the *universal* and *existential quantifier symbols*. A *quantifier* consists of a quantifier symbol and a bound variable symbol. For instance, $\forall \mathrm{x}$ and $\forall \mathrm{y}$ are *universal quantifiers*, $\exists \mathrm{x}$ and $\exists \mathrm{z}$ are *existential quantifiers*.

$\forall \mathrm{x}$ is the universal quantifier of x and is read as "for all (values of) x (in the domain)". $\exists \mathrm{x}$ is the existential quantifier of x and is read as " there exists some (value of) x (in the domain) such that". x is the bound variable symbol for quantification.

The last class includes three punctuation symbols, or simply, punctuation:

$$(\quad) \quad ,$$

which are the left and right parentheses and the *comma*.

When associated with a structure, $\mathcal{L}$ consists of the same classes of symbols as above, but the three classes of individual symbols, relation symbols, and function symbols should be in one-one correspondence with the designated individuals, relations, and functions of the structure; with n-ary relation symbols and function symbols corresponding to n-ary relations and functions. Therefore, these classes of symbols (excepting $\approx$) may differ from one first-order language to another, and are usually called *non-logical symbols*. The other classes of symbols (free and bound variable symbols, connectives, quantifier symbols, and punctuation) and $\approx$ are called *logical symbols.* They are assumed to be the same in all first-order languages (or, in the case of $\approx$, in all first-order languages with equality).

For convenience, a first-order language may be regarded as consisting of the non-logical symbols only. The equality symbol may or may not be included.

For instance, suppose $\mathcal{L}(\mathcal{N})$ is the first-order language associated with the structure $\mathcal{N} = \langle N, 0, =, ', +, \cdot \rangle$. The non-logical symbols of $\mathcal{L}(\mathcal{N})$ include an individual symbol a corresponding to 0, the equality symbol $\approx$ corresponding to $=$, a unary function symbol f corresponding to $'$, and two binary function symbols g and h corresponding to $+$ and $\cdot$. Then we have

$$\mathcal{L}(\mathcal{N}) = \langle \mathrm{a}, \approx, \mathrm{f}, \mathrm{g}, \mathrm{h} \rangle.$$

If we write a, f, g, h as 0, $'$, $+$, $\cdot$ respectively, we obtain

$$\mathcal{L}(\mathcal{N}) = \langle 0, \approx, ', +, \cdot \rangle.$$

However, it should be noticed that 0, $'$, $+$, $\cdot$ in $\mathcal{L}(\mathcal{N})$ are individual and function symbols, while those in $\mathcal{N}$ are individual and functions.

The first-order language $\mathcal{L}(\mathcal{G})$ associated with the structure $\mathcal{G}$ includes an individual symbol e corresponding to the unit element e of G, the equality symbol $\approx$, and a binary function symbol $\cdot$ corresponding to multiplication. Thus we have

$$\mathcal{L}(\mathcal{G}) = \langle \mathrm{e}, \approx, \cdot \rangle.$$

There is no restriction on the size of the set of non-logical symbols of a first-order language. It usually is finite or countably infinite.

From the expressions of $\mathcal{L}$, the sets of terms, atoms, and formulas of $\mathcal{L}$ are to be defined. They are denoted respectively by $Term(\mathcal{L})$, $Atom(\mathcal{L})$, and $Form(\mathcal{L})$.

Definition 3.2.1. (***Term***($\mathcal{L}$))

Term($\mathcal{L}$) is the smallest class of expressions of $\mathcal{L}$ closed under the following formation rules [1] and [2] of terms:

[1] a, u $\in$ *Term* ($\mathcal{L}$) (that is, an individual symbol or free variable symbol standing alone is a term of$\mathcal{L}$).

[2] If $\mathrm{t}_1, \ldots, \mathrm{t}_n \in$ *Term*($\mathcal{L}$) and f is an n-ary function symbol, then $\mathrm{f}(\mathrm{t}_1, \ldots, \mathrm{t}_n) \in$ Term($\mathcal{L}$).

Example

The expression f(g(f(a),g(u,b))) is a term of $\mathcal{L}$. It is left to the reader to see how it is formed.

In the languages $\mathcal{L}(\mathcal{N})$ and $\mathcal{L}(\mathcal{G})$, the expressions

$$+(\cdot('(0), \mathrm{u}), \ '('(0)))$$

and

$$\cdot(\mathrm{e}, \cdot(\mathrm{u}, \mathrm{e}))$$

are terms. They can be abbreviated to

$$0' \cdot \mathrm{u} + 0''$$

and

$$\mathrm{e} \cdot (\mathrm{u} \cdot \mathrm{e})$$

respectively, if $'(\mathrm{t})$, $+(\mathrm{t}_1, \mathrm{t}_2)$, and $\cdot(\mathrm{t}_1, \mathrm{t}_2)$ are abbreviated respectively as t', $\mathrm{t}_1 + \mathrm{t}_2$, and $\mathrm{t}_1 \cdot \mathrm{t}_2$ (t, t_1, and t_2 being terms).

The expression obtained at each step in the generation of terms is a term, and is a segment of the term generated.

The definition of *Term* ($\mathcal{L}$) is an inductive one.

The roman-type small Latin letter

$$\mathrm{t}$$

(with or without subscripts or superscripts) is used for any term.

Terms containing no free variable symbols are called *closed* terms. Thus

$$\mathrm{a}, \ \mathrm{f}(\mathrm{b}), \ \mathrm{g}(\mathrm{a}, \mathrm{f}(\mathrm{b}))$$

are closed terms, and

$$\mathrm{u}, \ \mathrm{g}(\mathrm{u}, \mathrm{b}), \ \mathrm{f}(\mathrm{g}(\mathrm{f}(\mathrm{u}), \mathrm{b}))$$

are not.

Inductive method can be used to prove any member of $Term(\mathcal{L})$ has certain property. The basis of induction is to prove any term containing no function symbol (that is, any individual symbol or free variable symbol) has this property. The induction step is to prove any term generated by means of function symbols preserves this property. This is called *a proof by induction on the structure (of generation) of terms.*

Some notational conventions will be stated for defining *Form* ($\mathcal{L}$). Suppose U, $V_1, \ldots, V_n$ are expressions and $s_1, \ldots, s_n$ are symbols of $\mathcal{L}$. We write

$$U(s_1, \ldots, s_n)$$

for U denoting that $s_1, \ldots, s_n$ occur in it. If $U(s_1, \ldots, s_n)$ occurs in the context, then

$$U(V_1, \ldots, V_n)$$

which follows is the expression resulting from $U(s_1, \ldots, s_n)$ by simultaneously substituting V_i for s_i in it $(i = 1, \ldots, n)$.

Example

Suppose U(a,u) = F(a) → G(a,u). Then U(u,a) = F(u) → G(u,a). U(u,a) is obtained from U(a,u) by simultaneously substituting u for a and a for u in it. If we substitute first u for a, obtaining F(u) → G(u,u), and then substitute a for u in this intermediate expression, we shall obtain F(a) → G(a,a), which is not correct.

Definition 3.2.2. (***Atom*** ($\mathcal{L}$))

An expression of $\mathcal{L}$ is a member of *Atom*($\mathcal{L}$) iff it is of one of the following two forms:

[1] $F(t_1, \ldots, t_n)$, where F is an n-ary relation symbol and $t_1, \ldots, t_n \in$ *Term*($\mathcal{L}$) $(n \geq 1)$.

[2] $\approx (t_1, t_2)$, where t_1, $t_2 \in$ *Term*($\mathcal{L}$).

We write $t_1 \approx t_2$ for $\approx (t_1, t_2)$.

Definition 3.2.3. (***Form***($\mathcal{L}$))

Form($\mathcal{L}$) is the smallest class of expressions of $\mathcal{L}$ closed under the following formation rules [1]–[4] of formulas of $\mathcal{L}$:

[1] *Atom*($\mathcal{L}$) $\subseteq$ *Form*($\mathcal{L}$).

[2] If A $\in$ *Form*($\mathcal{L}$), then $(\neg A) \in$ *Form*($\mathcal{L}$).

[3] If A, B $\in$ *Form*($\mathcal{L}$), then $(A * B) \in$ *Form*($\mathcal{L}$), $*$ being any one of $\wedge$, $\vee$, $\rightarrow$, and $\leftrightarrow$.

[4] If A(u) $\in$ *Form*($\mathcal{L}$), x not occurring in A(u), then $\forall$ xA(x), $\exists$ xA(x) $\in$ *Form*($\mathcal{L}$).

Example

The formula $\forall x(F(b) \rightarrow \exists y(\forall zG(y,z) \vee H(u,x,y)))$ is generated as follows:

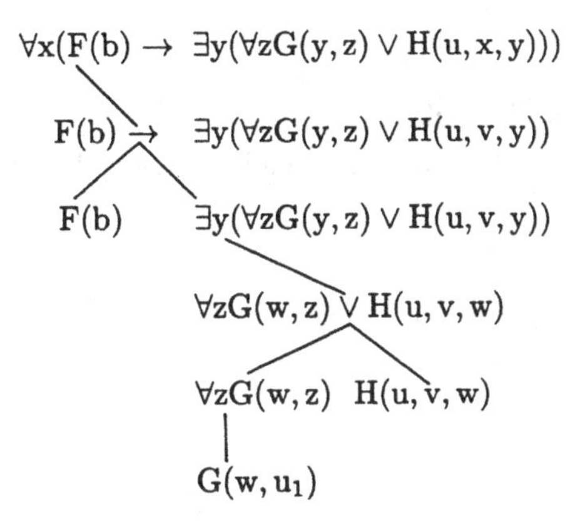

Remarks

(1) The formation rule [1] in Definition 3.2.3 of $Form(\mathcal{L})$ corresponds to [1] in Definition 2.2.2 of $Form(\mathcal{L}^p)$. Neither connectives nor quantifiers occur in atoms. [2] and [3] in Definition 3.2.3 are the same as those in Definition 2.2.2. [4] is used to generate new formulas by means of quantification. It is obvious from [4] that free variable symbols cannot be used together with quantifier symbols and a bound variable symbol x occurs in a formula iff $\forall x$ or $\exists x$ occurs in it.

(2) In generating a formula, we obtain at each step a formula which is not necessarily a segment of the formula generated because rule [4] requires substituting bound variable symbols for free ones. This is not the case with $Form(\mathcal{L}^p)$.

(3) By the condition " x not occurring in A(u)" in rule [4], $\exists x\forall yF(x,y)$ and $\forall x\forall yF(x,y)$ can be generated from $\forall yF(u,y)$, but neither $\exists y\forall yF(y,y)$ nor $\forall y\forall yF(y,y)$ can be generated. Therefore the x and y in $\exists x\forall yF(x,y)$ and $\forall x\forall yF(x,y)$ must be distinct bound variable symbols, while those in $\exists xG(x) \vee \forall yH(y)$ may or may not be distinct.

(4) Since it is stipulated in [4] that x does not occur in A(u), we see that A(u) contains u but not x, A(x) contains x but not u, and the occurrences of u in A(u) and those of x in A(x) correspond to each other. Besides the occurrences of u and x, the symbols of A(u) and A(x) are exactly the same. If it is not stipulated in [4] that x does not occur in A(u), then the occurrences of x in A(x) do not necessarily correspond to those of u in A(u).

The definition of $Form(\mathcal{L})$ is an inductive one.

The roman-type capital Latin letters

$$\mathrm{A} \quad \mathrm{B} \quad \mathrm{C}$$

are also used for any formula of $\mathcal{L}$.

Formulas with no free variable symbols are called *closed* formulas or *sentences*. Thus

$$\mathrm{F(a, b)}, \ \forall \mathrm{yF(a,y)}, \ \exists \mathrm{x} \forall \mathrm{yF(x,y)}$$

are sentences, while

$$\mathrm{F(u,v)}, \ \exists \mathrm{yF(u,y)}$$

are not.

The set of sentences of $\mathcal{L}$ is denoted by $Sent(\mathcal{L})$.

$Term(\mathcal{L})$, $Form(\mathcal{L})$, and $Sent(\mathcal{L})$ are countably infinite.

By Definition 3.2.3, ∀xA(x) and ∃xA(x) generated from A(u) are formulas. But the segment A(x) from them is not a formula because it contains x without any quantifier of x. A(x) is an expression resulting from A(u) by substituting x for u in it. The structural difference between such an expression and a formula lies in that the expression contains some bound variable symbol without any quantifier of it. Such expressions are called *quasi-formulas.* Quasi-formuals are also denoted by A, B, C.

Inductive method can be used to prove any formula of $\mathcal{L}$ has certain property. The basis of induction is to prove any atomic formula has this property. The induction step is to prove any formula generated by means of connectives or quantifiers preserves this property. This is called *a proof by induction on the structure (of generation) of formulas.*

Since the connectives of $\mathcal{L}^p$ are contained in $\mathcal{L}$, and the semantics and formal deduction of propositional logic are contained in first-order logic (see Sections 3.3–3.5), propositional logic is usually regarded as a part of first-order logic. This does not mean that propositional logic is completely contained in first-order logic. For instance, $\mathcal{L}$ does not contain proposition symbols. We may add proposition symbols into $\mathcal{L}$ and stipulate that proposition symbols are atoms of $\mathcal{L}$. Then $\mathcal{L}^p \subseteq \mathcal{L}$ and $Form(\mathcal{L}^p) \subseteq Form(\mathcal{L})$. But $\mathcal{L}$ has its own atoms, hence this is not necessary.

Propositions can be translated into formulas of $\mathcal{L}$. For instance, the proposition,

> For any natural number, there is a prime number
> greater than it.

can be restated as:

> For every x, if x is a natural number, then
> there is some y such that y is a prime number
> and $y > x$.

Suppose F and G are unary relation symbols and H is a binary relation symbol such that F(x) means " x is a natural number", G(x) means " x is a prime number", and H(x,y) means " x is greater than y". Then the above proposition is translated into the formula

$$\forall x[F(x) \to \exists y(G(y) \wedge H(y,x))].$$

Now we state without proof some structural properties of terms and formulas of $\mathcal{L}$, which are analogous to those discussed in Section 2.3 of Chapter 2.

Theorem 3.2.4.
Any term is of exactly one of three forms: an individual symbol, a free variable symbol or $f(t_1, \ldots, t_n)$, where f is an n-ary function symbol; and in each case it is of that form in exactly one way. □

Theorem 3.2.5.
If t is a segment of $f(t_1, \ldots, t_n)$, then t is a segment of any t_i ($i \doteq 1, \ldots, n$) or $t = f(t_1, \ldots, t_n)$. □

Theorem 3.2.6.
Any formula of $\mathcal{L}$ is of exactly one of eight forms: an atom, $(\neg A)$, $(A \wedge B)$, $(A \vee B)$, $(A \to B)$, $(A \leftrightarrow B)$, $\forall x A(x)$ or $\exists x A(x)$; and in each case it is of that form in exactly one way. □

Definition 3.2.7. (***Universal, existential formula***)
$\forall$ xA(x) is called a *universal formula*. It is the universal formula of A(u), u not occurring in A(x).

$\exists xA(x)$ is called an *existential formula*. It is the existential formula of A(u), u not occurring in A(x).

Definition 3.2.8. (***Scope***)

If $\forall\, xA(x)$ or $\exists xA(x)$ is a segment of B, A(x) is called the *scope* in B of the $\forall x$ or $\exists x$ on the left of A(x).

Theorem 3.2.9.

Any universal or existential quantifier in any formula has a unique scope. □

Theorem 3.2.10.

If A is a segment of $\forall\, xB(x)$ or $\exists\, xB(x)$, then A is a segment of B(x) or $A = \forall xB(x)$ or $\exists xB(x)$. □

It is obvious that the scope of a quantifier is not a formula but a quasi-formula, and if the scopes of a connective occur in the scope of a quantifier, then they may be quasi-formulas.

Example

In the formula $\exists x\forall y\exists zF(x,y,z)$, the scope of $\exists x$ is $\forall y\exists zF(x,y,z)$, that of $\forall y$ is $\exists zF(x,y,z)$, and that of $\exists z$ is $F(x,y,z)$. These scopes are all quasi-formulas.

In the formula $\exists x(G(u) \rightarrow H(u,x))$, the left scope of $\rightarrow$ is a formula G(u), while the right scope is a quasi-formula H(u,x).

The algorithms for deciding whether an expression of $\mathcal{L}$ belongs to *Form*($\mathcal{L}$) are omitted.

Exercises 3.2.

3.2.1. Translate the following propositions into formulas of $\mathcal{L}$ (select suitable symbols):

[1] All rational numbers are real numbers.

[2] All real numbers are not rational numbers.

[3] Some real numbers are not rational numbers.

[4] Not all real numbers are rational numbers.

[5] Every number is either odd or even.

[6] No number is both odd and even.
[7] 5 is divisible only by 1 and 5.
[8] If some trains are late, then all trains are late.

3.2.2. Suppose F(x, y) means "x likes y". Translate the following propositions into formulas of $\mathcal{L}$.
[1] Someone likes everyone.
[2] None likes everyone.

3.2.3. Suppose a denotes somebody, F(x) means "x is a job", and G(x, y) means "x can do y right". Translate:
[1] He can't do every job right.
[2] He can't do any job right.

3.2.4. Suppose F(x) means "x is a person", G(x) means "x is a time", and H(x, y) means "one can fool x at y". Translate (if the proposition is ambiguous, you will need more than one translation):
[1] One can fool some of the people at all of the time.
[2] One can fool all of the people at some of the time.
[3] One can't fool all of the people at all of the time.

3.2.5. Suppose F(x) means "x is a number", G(x) means "x is prime", H(x, y) means "x is less than y", and 0 denotes zero. Translate (if ambiguous, you will need more than one translation):
[1] Zero is less than any number.
[2] If any number is prime, than zero is prime.
[3] No number is less than zero.
[4] Any non-prime number with the property that all smaller numbers are prime is prime.
[5] There is no number such that all numbers are less than it.
[6] There is no number such that no number is less than it.

3.3. SEMANTICS

The first-order language $\mathcal{L}$, even though associated with a structure, is a syntactic object of no semantic significance. The formulas of $\mathcal{L}$, however, are intended to express propositions. This is accomplished by interpretations.

Interpretations for the propositional language are simple. They consist of assigning values to the proposition symbols. The first-order language includes more classes of symbols and hence the interpretations for it are more complicated.

Of the logical symbols, the connectives will be interpreted as in Chapter 2. The meaning of quantifiers has been explained intuitively in the last section. The equality symbol denotes the relation of equality. Free variable symbols will be interpreted as variables ranging over the domain. Punctuation symbols serve just like punctuation in natural languages.

The non-logical symbols, in the case where $\mathcal{L}$ is associated with a structure, are interpreted as the designated individuals, relations, and functions of the structure, with which they are in one-one correspondence. Accordingly, the sentences (closed formulas) are intended to express propositions about the structure.

If $\mathcal{L}$ is not associated with any structure, a domain is still required for each interpretation. In such cases, however, the domain is merely an arbitrary non-empty set. Then individual symbol, (n-ary) relation symbol, and (m-ary) function symbol are respectively interpreted as any individual in the domain, any (n-ary) relation and (m-ary) total function on the domain. Note that in such cases, different symbols of the same kind may have different or the same interpretation.

But it should be emphasized that the binary equality symbol is always interpreted as the equality relation on the domain.

To sum up, an interpretation for $\mathcal{L}$ consists of a domain and a function, which maps individual symbols, (n-ary) relation symbols, and (m-ary) function symbols respectively to individuals in the domain, (n-ary) relations and (m-ary) total functions on the domain. This is an interpretation for $\mathcal{L}$ in that domain.

We stipulate further, if an n-ary relation symbol F is interpreted as an n-ary relation R on a domain D, and terms $\mathrm{t}_1, \ldots, \mathrm{t}_n$ are respectively interpreted as individuals $\alpha_1, \ldots, \alpha_n$ in D, then the atomic formula

$$\mathrm{F}(\mathrm{t}_1, \ldots, \mathrm{t}_n)$$

is interpreted as the proposition

$$\alpha_1, \ldots, \alpha_n \text{ are in relation } R.$$

If an m-ary function symbol f is interpreted as an m-ary total function f on D, and terms $\mathrm{t}_1, \ldots, \mathrm{t}_m$ are respectively interpreted as $\alpha_1, \ldots, \alpha_m$ in

D, then the term

$$\mathrm{f}(\mathrm{t}_1, \ldots, \mathrm{t}_m)$$

is interpreted as the individual

$$f(\alpha_1, \ldots, \alpha_m)$$

in D.

Let N be the domain, the individual symbols a, b, and c in the closed term

$$\mathrm{f(g(a),f(b,c))}$$

are respectively interpreted as 4, 5 and 6, the binary and unary function symbols f and g are respectively interpreted as addition and squaring, then the above closed term is interpreted as

$$4^2 + (5 + 6)$$

which is the individual 27 in N.

Let N be the domain, and the interpretations of the symbols in the closed formula(sentence)

$$\mathrm{f(g(a),g(c))} \approx \mathrm{g(b)}$$

are the same as in the above example. Then the above closed formula is interpreted as the false proposition

$$4^2 + 6^2 = 5^2 .$$

However, the cases with non-closed terms and non-closed formulas are quite different. We first consider the case of terms. Let N be the domain, the interpretations of b, f and g in the term

1) $$\mathrm{f(g(u)),f(b,w))}$$

are the same as in the above example. Then 1) is interpreted as

2) $$x^2 + (5 + y)$$

where x and y are free variables with range N. Since 1) contains free variable symbols, hence 2) contains free variables, such that 2) is not an individual in N, but a binary function on N. Assigning individuals in N to x and y, we obtain the value of 2) for x and y at these individuals. It is called the value of 1) under the above interpretation together with the assignment of certain individuals in N to u and w in 1).

In general, a term containing m different free variable symbols is interpreted as an m-ary function on the domain. By interpretation together with an assignment of individuals in the domain to the free variable symbols, we obtain an individual in the domain as the value of the term.

Now we turn to the case of formulas. Suppose N be the domain. By the interpretation in the above example, the formula

3) $$f(g(u),g(w)) \approx g(b)$$

is interpreted as

4) $$x^2 + y^2 = 5^2$$

which is not a proposition, but a binary proposition function on N. Assigning individuals in N to x and y, we obtain a true or false proposition as the truth value of 4) for x and y at these individuals. It is called the truth value of 3) under the above interpretation together with the assignment of certain individuals in N to u and w in 3).

In general, a formula containing n different free variable symbols is interpreted as an n-ary proposition function on the domain. By interpretation together with an assignment of individuals in the domain to the free variable symbols, we obtain truth or falsehood as the truth value of the formula.

By the above explanations, we should distinguish between the interpretation of individual symbols as individuals in a domain and the assignment of indivduals to free variable symbols. The value of terms and the truth value of formulas of $\mathcal{L}$ depend not only upon interpretation, but also upon assignment of individuals to free variable symbols contained in terms or formulas. Thus, to obtain the value of terms and the truth value of formulas, we need an interpretation plus such an assignment.

Here there is a slight technical problem: different terms or formulas may involve different free variable symbols, so that we would have to consider an assignment of individuals to one set of free variable symbols in connection with one term or formula, and an assignment of individuals to a different set of free variable symbols in connection with another term or formula. This is feasible, but technically not convenient. Instead, we prefer to work with assignments that assign an individual in the domain to all free variable symbols at once. (Different or same individuals may be assigned to different free variable symbols.) And we shall arrange matters so that in evaluating any given term or formula the individuals assigned to free variable symbols which the term or formula does not involve will not in fact make any difference.

An interpretation together with an assignment is called a valuation, the domain of which is the same as that of the interpretation. We define valuation as follows. The italic small Latin letter v (with or without subscripts or superscripts) will be used for any valuation.

We recall (see Section 1.1) that an n-ary relation on domain D is a subset of D^n, the binary equality relation is

$$\{\langle x, y\rangle | x, y \in D \ \text{ and } \ x = y\}$$

or

$$\{\langle x, x\rangle | x \in D\}$$

which is a subset of D^2.

Definition 3.3.1. (***Valuation***)

A *valuation* v for the first-order language $\mathcal{L}$ consists of a domain D and a function (denoted by v) with the set of all individual symbols, relation symbols, function symbols, and free variable symbols as domain such that, writing a^v, F^v, $\approx^v$, f^v, and u^v respectively for $v(\mathrm{a})$, $v(\mathrm{F})$, $v(\approx)$, $v(\mathrm{f})$ and $v(\mathrm{u})$ (a, F, f and u being respectively any individual symbol, n-ary relation symbol, m-ary function symbol, and free variable symbol), we have

[1] $\mathrm{a}^v, \mathrm{u}^v \in D$.

[2] $\mathrm{F}^v \subseteq D^n$;

$\approx^v = \{\langle x, x\rangle | x \in D\} \subseteq D^2$.

[3] $\mathrm{f}^v : D^m \to D$.

Remarks

(1) a, F, $\approx$, f, and u should be distinguished respectively from a^v, F^v, $\approx^v$, f^v, and u^v. The former are symbols in $\mathcal{L}$, while the latter are the valuation which v gives to the symbols.

a^v, F^v, f^v, and u^v are determined by v. The valuation caused by v changes with the domain of v. Even if the domain remains unchanged, a^v, F^v, f^v, and u^v change with v. But the case with $\approx$ is quite different. When the domain of v remains unchanged, $\approx^v$ is always the equality relation on this domain. $\approx^v$ becomes equality relation on another domain, only when the domain changes.

(2) The domain of f^v is D^m; that is, f^v is an m-ary total function on D.

(3) We mentioned before that we should distinguish between the interpretation of individual symbols as individuals and the assignment of individuals to free variable symbols. Hence we cannot regard individual symbols and free variable symbols as of the same kind from $a^v \in D$ and $u^v \in D$ in Definition 3.3.1.

The value of a term t under a valuation v is denoted by t^v. The truth value of a formula A under v is denoted by A^v.

Definition 3.3.2. (***Value of Terms***)

The *value of terms* of $\mathcal{L}$ under valuation v over domain D is defined by recursion:

[1] a^v, $u^v \in D$.

[2] $f(t_1, \ldots, t_n)^v = f^v(t_1^v, \ldots, t_n^v)$.

Theorem 3.3.3.

Suppose v is a valuation over domain D, and $t \in Term(\mathcal{L})$. Then $t^v \in D$.

Proof. By induction on the structure of t. □

To define the value of formulas under valuations, we introduce the following notational convention. Suppose $\alpha \in D$. We write

$$v(u/\alpha)$$

for a valuation which is exactly the same as v except that $u^{v(u/\alpha)} = \alpha$. That is, for any individual symbol a, relation symbol F, function symbol f, and free variable symbol w, we have

$$a^{v(u/\alpha)} = a^v$$
$$F^{v(u/\alpha)} = F^v$$
$$f^{v(u/\alpha)} = f^v$$
$$w^{v(u/\alpha)} = \begin{cases} \alpha & \text{if } w = u, \\ w^v & \text{otherwise.} \end{cases}$$

Definition 3.3.4. (***Truth value of formulas***)

The *truth value of formulas* of $\mathcal{L}$ under valuation v over domain D is defined by recursion:

[1] $F(t_1,\ldots,t_n)^v = \begin{cases} 1 & \text{if } \langle t_1^v,\ldots,t_n^v\rangle \in F^v, \\ 0 & \text{otherwise.} \end{cases}$

$(t_1, \approx t_2)^v = \begin{cases} 1 & \text{if } t_1^v = t_2^v, \\ 0 & \text{otherwise.} \end{cases}$

[2] $(\neg A)^v = \begin{cases} 1 & \text{if } A^v = 0, \\ 0 & \text{otherwise.} \end{cases}$

[3] $(A \wedge B)^v = \begin{cases} 1 & \text{if } A^v = B^v = 1, \\ 0 & \text{otherwise.} \end{cases}$

[4] $(A \vee B)^v = \begin{cases} 1 & \text{if } A^v = 1 \text{ or } B^v = 1, \\ 0 & \text{otherwise.} \end{cases}$

[5] $(A \rightarrow B)^v = \begin{cases} 1 & \text{if } A^v = 0 \text{ or } B^v = 1, \\ 0 & \text{otherwise.} \end{cases}$

[6] $(A \leftrightarrow B)^v = \begin{cases} 1 & \text{if } A^v = B^v, \\ 0 & \text{otherwise.} \end{cases}$

[7] $\forall x A(x)^v = \begin{cases} 1 & \text{if, constructing } A(u) \text{ from } A(x) \text{ (taking u not occurring in } A(x)), \text{ for every } \alpha \in D, \ A(u)^{v(u/\alpha)} = 1, \\ 0 & \text{otherwise.} \end{cases}$

[8] $\exists x A(x)^v = \begin{cases} 1 & \text{if, constructing } A(u) \text{ from } A(x) \text{ (taking u not occurring in } A(x)), \text{ there exists } \alpha \in D, \text{ such that } A(u)^{v(u/\alpha)} = 1, \\ 0 & \text{otherwise.} \end{cases}$

Remarks

(1) "$\langle t_1^v,\ldots,t_n^v\rangle \in F^v$" in Definition 3.3.4 [1] means that $t_1^v,\ldots,t_n^v$ are in relation F^v; "$t_1^v = t_2^v$" means that t_1^v and t_2^v are in equality relation $\approx^v$.

(2) In Definition 3.3.4 [7] and [8], it is stipulated that, in constructing A(u), we take free variable symbol u not occurring in A(x). This is feasible, since A(x) is of finite length and free variable symbols are countably infinite in number.

Besides, according to the construction of A(u), x does not occur in A(u). As explained before, the occurrences of x in A(x) and those of u in A(u) correspond to each other, and the symbols of A(x) and A(u) other than the occurrences of x and u are exactly the same. Hence A(x) and A(u) have the same intuitive meaning; that is, A(x) talks about x in exactly the same way as A(u) talks about u. But the case with u occurring in A(x) is quite different.

(3) $\forall$xA(x) and $\exists$xA(x) are generated from A(u), hence in the recursive Definition 3.3.4, $\forall xA(x)^v$ and $\exists xA(x)^v$ are defined from $A(u)^v$. Intuitively speaking, if the proposition $A(u)^v$ means the individual u^v in some domain D has certain property, then the proposition $\forall xA(x)^v$ means every individual in D has this property, and $\exists xA(x)^v$ means some individual (or individuals) in D has this property. Hence, $\forall xA(x)^v = 1$ means that, no matter what individual u^v is in D (that is, no matter what individual in D v assigns to u as its value), we have $A(u)^v = 1$. $\exists xA(x)^v = 1$ means that there exists some individual in D such that, supposing u^v is this individual (that is, supposing v assigns this individual to u as its value), we have $A(u)^v = 1$.

(4) $\forall$xA(x) and $\exists$xA(x) may contain occurrences of free variable symbols, hence in addition to the free variable symbol u (which is used in constructing A(u)), A(u) may contain occurrences of free variable symbols, say w (which occur originally in $\forall$xA(x) or $\exists$xA(x)). As explained in (3), in obtaining $A(u)^v = 1$ in the case of $\forall xA(x)^v = 1$, the assignment u^v needs to cover the whole domain D, while in the case of $\exists xA(x)^v = 1$, the assignment u^v is to cover a part of D. But now the problem is, for each w (if any) in A(u), w^v should be the same as v assign to w in $\forall$xA(x) or $\exists$xA(x). In order to express this precisely, we use the valuation $v(u/\alpha)$ to replace the original valuation v, and require

$$\text{For every } \alpha \in D,\ A(u)^{v(u/\alpha)}{=}1.$$

and

$$\text{There exists some } \alpha \in D, \text{ such that } A(u)^{v(u/\alpha)} = 1.$$

respectively in [7] and [8].

(5) Valuation $v(u/\alpha)$ is used for evaluating $\forall xA(x)^v$ and $\exists xA(x)^v$. As explained before, in evaluating any given term or formula the individuals assigned to free variable symbols which the term or formula does not involve will not in fact make any difference. Hence, although assignments are stipulated to assign simultaneously an individual to all free variable symbols, yet for a given term or formula, a valuation may be regarded to assign individuals only to the free variable symbols involved in it. Thus, if v is to assign to the free variable symbols in $\forall$xA(x) or $\exists$xA(x), then $v(u/\alpha)$ is to assign not only to these free variable symbols, but also to the free variable symbol u, which does not occur in $\forall$xA(x) or $\exists$xA(x).

(6) The "otherwise" in Definition 3.3.4 [7] means there exists some $\alpha \in D$, such that $A(u)^{v(u/\alpha)} = 0$, and the "otherwise" in [8] means for every $\alpha \in D$, $A(u)^{v(u/\alpha)} = 0$ (constructing A(u) from A(x) and taking u not occurring in A(x)).

Theorem 3.3.5.

Suppose v is a valuation over domain D and $A \in Form(\mathcal{L})$. Then $A^v \in \{1, 0\}$.

Proof. By induction on the structure of A. □

We note that valuations are analogous to but not the same as truth valuations defined in Section 2.4.

In evaluating a term t or a formula A under valuation v, we need only the finite amount of information concerning a^v, F^v, f^v, and u^v, where a, F, f, and u denote the non-logical symbols and free variable symbols occurring in t or A.

Two valuations v and v' over the same domain are said to agree on non-logical symbol a (or F, f) or free variable symbol u if $a^v = a^{v'}$ (or $F^v = F^{v'}$, $f^v = f^{v'}$) or $u^v = u^{v'}$.

Theorem 3.3.6.

Suppose v and v' are two valuations over the same domain, and they agree on the individual symbols, relation symbols, function symbols, and free variable symbols contained in term t and formula A. Then

[1] $t^v = t^{v'}$.

[2] $A^v = A^{v'}$.

Proof. By induction on the structures of t and A. □

Suppose $\Sigma \subseteq Form(\mathcal{L})$. We define

$$\Sigma^v = \begin{cases} 1 & \text{if for every } B \in \Sigma, B^v = 1, \\ 0 & \text{otherwise.} \end{cases}$$

Definition 3.3.7. ***(Satisfiability)***

$\Sigma \subseteq Form(\mathcal{L})$ is *satisfiable* iff there is some valuation v such that $\Sigma^v = 1$.

When $\Sigma^v = 1$, we say v *satisfies* Σ.

Definition 3.3.8. ***(Validity)***

$A \in Form(\mathcal{L})$ is *valid* iff for every valuation v, $A^v = 1$.

Validity is also called *universal validity.*

A valid formula is one that is true on account of its form alone, irrespective of the meaning of the non-logical symbols and free variable symbols yielded under interpretations and assignments. Validity is intended to capture the informal notion of truth of proposition with attention to the logical form in abstraction from the matter.

A satisfiable formula (or set of formulas) is one that is true relative to some particular interpretation and assignment. Hence satisfiability corresponds to the informal notion of truth of propositions which follows from the matter.

Satisfiability and validity are important semantic notions which are closely related to each other. They will be studied further in Chapter 5.

Example

Suppose $A = f(g(a), g(u)) \approx g(b)$, v is a valuation over domain N such that $a^v = 3$, $b^v = 5$, $u^v = 4$, f^v is addition, and g^v is squaring. Then A^v is the true proposition

(1) $$3^2 + 4^2 = 5^2.$$

Hence A is satisfiable. The truth of (1) is determined by the matter. In fact, there are other valuations which make A true. But A is not valid. If we set $b^v = 6$ in the above valuation, A^v will be false.

Suppose $B = F(u) \vee \neg F(u)$, v is any valuation. Then, B^v is the true proposition

(2) $$u^v \text{ has or has not the property } F^v.$$

The truth of (2) is not concerned with the domain, the individual u^v, or the property F^v. It follows from the logical form which justifies the validity of B.

Valid formulas in $\mathcal{L}$ are the counterpart of tautologies in $\mathcal{L}^p$. The similarities between them are obvious, but there is one important difference. To decide whether a formula of $\mathcal{L}^p$ is a tautology, algorithms are used (for instance, the truth table). However, in order to know whether a formula of $\mathcal{L}$ is valid, we have to consider all valuations over domains of different sizes. In the case of an infinite domain D, the procedure is in general not finitary. We are not provided with a method for evaluating $\forall x A(x)^v$ or $\exists x A(x)^v$ in a finite number of steps, because it presupposes the values $A(u)^{v(u/\alpha)}$ for infinitely many $\alpha \in D$. (See Definition 3.3.4.)

It is sometimes possible to decide for certain formulas of $\mathcal{L}$ whether they are valid or not. However, Church [1936] established that there is no algorithm for deciding the validity or satisfiability of formulas of $\mathcal{L}$. This topic belongs to recursion theory, a branch of mathematical logic not contained in this book.

3.4. LOGICAL CONSEQUENCE

Logical consequences in $\mathcal{L}$, which are the counterpart of tautological consequences in $\mathcal{L}^p$, involve semantics. The notation $\models$ for tautological consequences is also used for logical consequences.

Definition 3.4.1. (***Logical consequence***)

Suppose $\Sigma \subseteq Form(\mathcal{L})$ and $A \in Form(\mathcal{L})$.

A is a *logical consequence* of Σ, written as $\Sigma \models A$, iff for any valuation v, $\Sigma^v = 1$ implies $A^v = 1$.

In the special case of $\emptyset \models A$, A is valid.

The notations $\not\models$ and $\models\!\!\!\!\dashv$ are used in the same way as in Chapter 2.

Two formulas A and B are called *logically equivalent* (or *equivalent* for short, if no confusion will arise) iff $A \models\!\!\!\!\dashv B$ holds.

We will show with examples how to prove or refute a logical consequence. This is analogous to the case of tautological consequences in Section 2.5.

Example

$\forall x \neg A(x) \models \neg \exists x A(x)$.

Proof. Suppose $\forall x \neg A(x) \not\models \neg \exists x A(x)$, that is, there is some valuation v over domain D such that

$$(1) \qquad \forall x \neg A(x)^v = 1;$$

$$(2) \qquad (\neg \exists x A(x))^v = 0.$$

Form A(u) from A(x), u not occurring in A(x). By (1) we obtain $(\neg A(u))^{v(u/\alpha)} = 1$ for every $\alpha \in D$, hence

$$(3) \qquad \text{For every } \alpha \in D, A(u)^{v(u/\alpha)} = 0.$$

By (2) we obtain $\exists x A(x)^v = 1$, contradicting (3). Hence $\forall x \neg A(x) \models \neg \exists x A(x)$.

Example

$\forall x(A(x) \to B(x)) \models \forall x A(x) \to \forall x B(x)$.

Proof. Suppose $\forall x(A(x) \to B(x)) \not\models \forall xA(x) \to \forall xB(x)$, that is, there is some valuation v over domain D such that

$$(1) \qquad \forall x(A(x) \to B(x))^v = 1;$$
$$(2) \qquad (\forall xA(x) \to \forall xB(x))^v = 0.$$

By (2) we have

$$(3) \qquad \forall xA(x)^v = 1;$$
$$(4) \qquad \forall xB(x)^v = 0.$$

Form A(u) and B(u), u not occurring in A(x) or in B(x). By (1), (3), and (4) we obtain respectively (5), (6) and (7):

$$(5) \qquad \text{For every } \alpha \in D,\ (A(u) \to B(u))^{v(u/\alpha)} = 1;$$
$$(6) \qquad \text{For every } \alpha \in D,\ A(u)^{v(u/\alpha)} = 1;$$
$$(7) \qquad \text{For some } \alpha \in D,\ B(u)^{v(u/\alpha)} = 0.$$

By (5) and (6) we obtain $B(u)^{v(u/\alpha)} = 1$ for every $\alpha \in D$, contradicting (7). Therefore $\forall x(A(x) \to B(x)) \models \forall xA(x) \to \forall xB(x)$.

In the above examples logical consequences are proved. We need not construct valuations in the proofs. In the following example a logical consequence is to be refuted. We need to construct a valuation in the refutation and determine its domain. We want to explain that what is to be determined about the domain is its cardinal (that is, its size), irrelevant to what members it contains. We explain it with example.

Suppose we want to construct a valuation v for the atomic formula F(u). Take the set $\{\alpha, \beta\}$ with two members as its domain. Then u^v may be 1) or 2):

$$1) \qquad u^v = \alpha;$$
$$2) \qquad u^v = \beta.$$

F^v may be one of 3)–6):

$$3) \qquad F^v = \{\alpha, \beta\};$$
$$4) \qquad F^v = \{\alpha\};$$
$$5) \qquad F^v = \{\beta\};$$
$$6) \qquad F^v = \emptyset.$$

Then, if 1) and 3) are used, we obtain $F(u)^v = 1$; if 1) and 5) are used, we obtain $F(u)^v = 0$, etc.

We may take another domain $\{\alpha', \beta'\}$, where α' and β' are different from α and β. Let α and β correspond to α' and β' respectively. We construct a valuation v' over domain $\{\alpha', \beta'\}$. Then $u^{v'}$ may be 1′) or 2′):

1) $$u^v = \alpha';$$
2) $$u^v = \beta'.$$

$F^{v'}$ may be one of 3′)–6′):

3′) $$F^{v'} = \{\alpha', \beta'\};$$
4′) $$F^{v'} = \{\alpha'\};$$
5′) $$F^{v'} = \{\beta'\};$$
6′) $$F^{v'} = \emptyset.$$

When 1′) and 3′) are used, $F(u)^{v'} = 1$; when 1′) and 5′) are used, $F(u)^{v'} = 0$, etc.

We see that the result of constructing a valuation for F(u) with domain $\{\alpha, \beta\}$ can be obtained with $\{\alpha', \beta'\}$. Thus we explain intuitively that the crucial point of the domain of valuation is its cardinal.

Example

$\forall xA(x) \to \forall xB(x) \not\models \forall x(A(x) \to B(x))$.

To refute a logical consequence, we need only to refute a special example of it. Suppose the quasi-formulas A(x) and B(x) are atomic quasi-formulas F(x) and G(x) respectively. Then we are to prove $\forall xF(x) \to \forall\, xG(x) \not\models \forall\, x[F(x) \to G(x)]$.

Proof. Set $D = \{\alpha, \beta\}$. Form F(u) and G(u). Construct a valuation v over domain D such that $F^v = \{\alpha\}$ and $G^v = \{\beta\}$ or $\emptyset$. The

(1) $$F(u)^{v(u/\alpha)} = 1;$$
(2) $$F(u)^{v(u/\beta)} = 0;$$
(3) $$G(u)^{v(u/\alpha)} = 0;$$

where $G(u)^{v(u/\beta)}$ is irrelevant to the question. Then we obtain the following:

(4) $\forall xF(x)^v = 0$ (by (2));

(5) $(\forall xF(x) \to \forall xG(x))^v = 1$ (by (4));

(6) $(F(u) \to G(u))^{v(u/\alpha)} = 0$ (by (1), (3));

(7) $\forall x(F(x) \to G(x))^v = 0$ (by (6)).

By (5) and (7) we obtain

$$\forall xF(x) \to \forall xG(x) \not\models \forall x(F(x) \to G(x)).$$

Note that the statement in this example cannot be proved in a domain with only one individual.

Lemma 3.4.2.
Suppose A $\models\!\dashv$ A′, B $\models\!\dashv$ B′ and C(u) $\models\!\dashv$ C′(u). Then

[1] $\neg A \models\!\dashv \neg A'$.
[2] $A \wedge B \models\!\dashv A' \wedge B'$.
[3] $A \vee B \models\!\dashv A' \vee B'$.
[4] $A \rightarrow B \models\!\dashv A' \rightarrow B'$.
[5] $A \leftrightarrow B \models\!\dashv A' \leftrightarrow B'$.
[6] $\forall x C(x) \models\!\dashv \forall x C'(x)$.
[7] $\exists x C(x) \models\!\dashv \exists x C'(x)$.

Proof. [1]–[5] are the same as those in Section 2.5. We shall prove [6], and [7] is left to the reader.

Suppose v is any valuation over domain$D$. Suppose

$$\forall x C(x)^v = 1. \tag{1}$$

Form C(u) and C′(u), u not occurring in C(x) or in C′(x). By (1) we obtain

$$\text{For every } \alpha \in D, C(u)^{v(u/\alpha)} = 1. \tag{2}$$

By (2) and the suppositions of the lemma, we obtain $C'(u)^{v(u/\alpha)} = 1$ for every $\alpha \in D$, and then $\forall x C'(x)^v = 1$. Hence $\forall x C(x) \models \forall x C'(x)$. Similarly for the proof of $\forall x C'(x) \models \forall x C(x)$. □

Suppose B and C are quasi-formulas, for instance, $B = F(x) \vee G(u, x)$ and $C = \neg F(x) \rightarrow G(u, x)$, x occurring in them without quantifiers of it. B $\models\!\dashv$ C is meaningless because B and C are not formulas. Substituting any free variable symbol v for x in B and C, we obtain formulas B′ and C′:

$$B' = F(v) \vee G(u, v)$$
$$C' = \neg F(v) \rightarrow G(u, v)$$

and B′ $\models\!\dashv$ C′. In such a case, B $\models\!\dashv$ C is intended to mean B′ $\models\!\dashv$ C′.

Generally, if B and C are quasi-formulas containing $x_1, \ldots, x_n$ without quantifiers of them and B′, C′ are formulas resulting from B, C by simultaneously substituting any free variable symbols $u_1, \ldots, u_n$ respectively for $x_1, \ldots, x_n$ in B and C, we write B $\models\!\dashv$ C for B′ $\models\!\dashv$ C′.

Theorem 3.4.3. (***Replaceability of equivalent formulas***)
Suppose B $\models\!\dashv$ C and A′ results from A by replacing some (not necessarily all) occurrences of B in A by C. Then A $\models\!\dashv$ A′.

Proof. By induction on the structure of A, using Lemma 3.4.2. □

Note that B and C in Theorem 3.4.3 may be quasi-formulas.

Theorem 3.4.4. (***Duality***)
Suppose A is a formula composed of atoms of $\mathcal{L}$, the connectives $\neg$, $\wedge$, $\vee$, and the two quantifiers by the formation rules concerned, and A′ (the *dual* of A) results from A by exchanging $\wedge$ for $\vee$, $\forall$ for $\exists$ and each atom for its negation. Then A′ $\models\!\dashv$ $\neg$A.

Proof. By induction on the structure of A. □

Exercises 3.4.

3.4.1. Prove Theorem 3.4.4.

3.4.2. Prove the following logical consequences:
[1] $\neg\forall xA(x) \models\!\dashv \exists x\neg A(x)$
[2] $\neg\exists xA(x) \models\!\dashv \forall x\neg A(x)$
[3] $\forall x(A(x) \wedge B(x)) \models\!\dashv \forall xA(x) \wedge \forall xB(x)$
[4] $\exists x(A(x) \vee B(x)) \models\!\dashv \exists xA(x) \vee \exists xB(x)$

3.4.3. Prove the following:
[1] $\exists x(A(x) \wedge B(x)) \models \exists xA(x) \wedge \exists xB(x)$
[2] $\exists xA(x) \wedge \exists xB(x) \not\models \exists x(A(x) \wedge B(x))$
[3] $\forall xA(x) \vee \forall xB(x) \models \forall x(A(x) \vee B(x))$
[4] $\forall x(A(x) \vee B(x)) \not\models \forall xA(x) \vee \forall xB(x)$
[5] $\exists x\forall yA(x,y) \models \forall y\exists xA(x,y)$
[6] $\forall y\exists xA(x,y) \not\models \exists x\forall yA(x,y)$

3.5. FORMAL DEDUCTION

The formal deduction in first-order logic is analogous to that in propositional logic except that it is defined by additional rules of formal deduction.

The eleven rules in propositional logic are included in first-order logic, but the formulas occurring in them are now formulas of the first-order language. The additional rules concerning the quantifiers and equality symbol are as follows:

($\forall-$) If $\Sigma \vdash \forall x A(x)$,
then $\Sigma \vdash A(t)$. (*$\forall$–elimination*)

($\forall+$) If $\Sigma \vdash A(u)$, u not occurring in Σ,
then $\Sigma \vdash \forall x A(x)$. (*$\forall$–introduction*)

($\exists-$) If $\Sigma, A(u) \vdash B$, u not occurring in Σ or B,
then $\Sigma, \exists x A(x) \vdash B$. (*$\exists$–elimination*)

($\exists+$) If $\Sigma \vdash A(t)$,
then $\Sigma \vdash \exists x A(x)$, where A(x) results by replacing some (not necessarily all) occurrences of t in A(t)by x.
(*$\exists$–introduction*)

($\approx-$) If $\Sigma \vdash A(t_1)$,
$\Sigma \vdash t_1 \approx t_2$,
then $\Sigma \vdash A(t_2)$, where $A(t_2)$ results by replacing some (not necessarily all) occurrences of t_1 in $A(t_1)$by t_2.
(*$\approx$–elimination*)

($\approx+$) $\emptyset \vdash u \approx u$. (*$\approx$ –introduction*)

Remarks

In ($\forall-$) the formula A(t) results from A(x) by substituting t for all occurrences of x. It is the same for the cases of ($\forall+$) and ($\exists-$). In ($\exists+$) and ($\approx-$), however, another kind of replacement is employed, which should be distinguished from substitution.

The u's in ($\forall+$) and ($\exists-$) may be replaced by t. Such replacement extends the range of application of these rules, because the set of terms contains the set of free variable symbols as a proper subset. However, since the above original formulations are sufficient, the replacement of u by t is not necessary.

Note that the t's in $(\forall -)$ and $(\exists +)$ should not be replaced by u.

The conditions "u not occurring in Σ" in $(\forall +)$ and "u not occurring in Σ or B" in $(\exists -)$ call for some explanation. The rule $(\forall +)$ means intuitively that from "any member α of a set has a certain property" we can deduce "every member of the set has this property". For instance, in order to demonstrate that every point on the perpendicular bisector of a segment AB is equidistant from A and B, it is sufficient to demonstrate this statement for any point α on the perpendicular bisector. The arbitrariness of α means that the choice of α is independent of the premises in the deduction. This point is expressed in $(\forall +)$ by "u not occurring in Σ", where u expresses α and Σ expresses the premises. This is similar for the case of $(\exists -)$. The following sequences:

$$\begin{cases} (1) \quad A(u) \vdash A(u) \quad (\text{by (Ref)}) \\ (2) \quad A(u) \vdash \forall x A(x) \quad (\text{by } (\forall +), (1)) \end{cases}$$

$$\begin{cases} (1) \quad A(u) \vdash A(u) \quad (\text{by (Ref)}) \\ (2) \quad \exists x A(x) \vdash A(u) \quad (\text{by } (\exists -), (1)) \end{cases}$$

do not form (formal) proofs because the rules $(\forall +)$ and $(\exists -)$ are used incorrectly in them. In fact, $A(u) \vdash \forall x A(x)$ and $\exists x A(x) \vdash A(u)$ do not hold. This can be proved after reading Soundness Theorem in Section 5.2.

Definition 3.5.1. (***Formal deducibility***)

Suppose $\Sigma \subseteq \mathit{Form}(\mathcal{L})$ and $A \in \mathit{Form}(\mathcal{L})$.

A is *formally deducible* from Σ in first-order logic iff $\Sigma \vdash A$ can be generated by the seventeen rules of formal deduction.

For a sequence of universal (or existential) quantifiers, we write

$$\forall x_1 \ldots x_n$$

$$\exists x_1 \ldots x_n$$

for $\forall x_1 \ldots \forall x_n$ and $\exists x_1 \ldots \exists x_n$ respectively.

Theorem 2.6.2 holds in classical first-order logic as well, the proof of which is left to the reader.

The rules of formal deduction for quantifiers can be generalized as follows:

($\forall-$) If $\Sigma \vdash \forall x_1 \ldots x_n A(x_1, \ldots, x_n)$,
then $\Sigma \vdash A(t_1, \ldots, t_n)$.

($\forall+$) If $\Sigma \vdash A(u_1, \ldots, u_n), u_1, \ldots, u_n$ not occurring in Σ,
then $\Sigma \vdash \forall x_1 \ldots x_n A(x_1, \ldots, x_n)$.

($\exists-$) If $\Sigma, A(u_1, \ldots, u_n) \vdash B, u_1, \ldots, u_n$ not occurring in Σ or B,
then $\Sigma, \exists x_1 \ldots x_n A(x_1, \ldots, x_n) \vdash B$.

($\exists+$) If $\Sigma \vdash A(t_1, \ldots, t_n)$,
then $\Sigma \vdash \exists x_1 \ldots x_n A(x_1, \ldots, x_n)$, where $A(x_1, \ldots, x_n)$ results by replacing simultaneously some (not necessarily all) occurrences of t_i in $A(t_1, \ldots, t_n)$ by x_i ($i = 1, \ldots, n$).

In the above generalizations, $x_1, \ldots, x_n$ should be distinct, otherwise we shall have, for instance, $\forall xx A(x, x)$ (that is, $\forall x \forall x A(x, x)$), which is not a formula. In ($\forall+$) and ($\exists-$), $u_1, \ldots, u_n$ should be distinct, otherwise we shall have the same symbols in $x_1, \ldots, x_n$, and accordingly $\forall x_1 \ldots x_n A(x_1, \ldots, x_n)$ and $\exists x_1 \ldots x_n A(x_1, \ldots, x_n)$ will not be formulas. But in ($\forall-$) and ($\exists+$), $t_1, \ldots, t_n$ may or may not be distinct. For instance, from

$$\forall xyz F(x, y, z) \vdash \forall xyz F(x, y, z)$$

we can have

$$\forall xyz F(x, y, z) \vdash F(t_1, t_2, t_3),$$
$$\forall xyz F(x, y, z) \vdash F(t_1, t_2, t_2),$$
$$\forall xyz F(x, y, z) \vdash F(t_1, t_1, t_1).$$

Obviously, every scheme of formal deducibility which holds in propositional logic holds in first-order logic as well, but the formulas occurring in it should be replaced by formulas of $\mathcal{L}$.

Theorem 3.5.2.

[1] $\forall x_1 \ldots x_n A(x_1, \ldots, x_n) \vdash A(t_1, \ldots, t_n)$.

[2] $A(t_1, \ldots, t_n) \vdash \exists x_1 \ldots x_n A(x_1, \ldots, x_n)$, where $A(x_1, \ldots, x_n)$ results by replacing simultaneously some (not necessarily all) occurrences of t_i in $A(t_1, \ldots, t_n)$ by x_i ($i = 1, \ldots, n$).

[3] $\forall x A(x) \vdash \forall y A(y)$.

[4] $\exists xA(x) \vdash \exists yA(y)$.
[5] $\forall xyA(x, y) \vdash \forall yxA(x, y)$.
[6] $\exists xyA(x, y) \vdash \exists yxA(x, y)$.
[7] $\forall xA(x) \vdash \exists xA(x)$.
[8] $\exists x\forall yA(x, y) \vdash \forall y\exists xA(x, y)$.

Proof. We choose to prove [4] and [5].

Proof of [4].
(1) $A(u) \vdash A(u)$ (use u not occurring in $A(y)$).
(2) $A(u) \vdash \exists yA(y)$ (by ($\exists+$), (1)).
(3) $\exists xA(x) \vdash \exists yA(y)$ (by ($\exists-$), (2)).

Proof of [5].
(1) $\forall xyA(x, y) \vdash A(u, v)$ (use u, v not occurring in $A(x, y)$).
(2) $\forall xyA(x, y) \vdash \forall yxA(x, y)$ (by ($\forall+$), (1)). □

Theorem 3.5.3.
[1] $\neg\forall xA(x) \vdash\dashv \exists x\neg A(x)$.
[2] $\neg\exists xA(x) \vdash\dashv \forall x\neg A(x)$.

Proof. We choose to prove [1].
Proof of $\vdash$ of [1].
(1) $\neg A(u) \vdash \exists x\neg A(x)$ (use u not occurring in $A(x)$).
(2) $\neg\exists x\neg A(x) \vdash A(u)$ (by (1)).
(3) $\neg\exists x\neg A(x) \vdash \forall xA(x)$.
(4) $\neg\forall xA(x) \vdash \exists x\neg A(x)$ (by (3)).
Proof of $\dashv$ of [1].
(1) $\forall xA(x) \vdash A(u)$ (use u not occurring in $A(x)$).
(2) $\neg A(u) \vdash \neg\forall xA(x)$ (by (1)).
(3) $\exists x\neg A(x) \vdash \neg\forall xA(x)$ (by ($\exists-$), (2)). □

Remarks

$\Sigma \vdash \exists xA(x)$, where $\exists xA(x)$ is an existential formula, may be proved from $\Sigma \vdash A(t)$ by using ($\exists+$). Such proof is called constructive. Another method is to obtain $\Sigma \vdash \exists xA(x)$ from

$$\Sigma, \neg\exists xA(x) \vdash B$$
$$\Sigma, \neg\exists xA(x) \vdash \neg B$$

by using $(\neg-)$. The above proof of the $\vdash$ part of [1] is non-constructive. (Constructive logic will be studied in Chapter 7.)

If we can establish $\neg\forall xA(x) \vdash \neg A(u)$, then we obtain $\neg\forall xA(x) \vdash \exists x\neg A(x)$ from it by using $(\exists+)$. This gives a constructive proof of the $\vdash$ part of [1]. But we have $\neg\forall xA(x) \not\models \neg A(u)$. Hence, by Soundness Theorem (see Section 5.2), $\neg\forall xA(x) \vdash \neg A(u)$ does not hold. In fact, the $\vdash$ part of [1] has no constructive proof.

Theorem 3.5.4.
[1] $\forall x(A(x) \to B(x)) \vdash \forall xA(x) \to \forall xB(x)$.
[2] $\forall x(A(x) \to B(x)) \vdash \exists xA(x) \to \exists xB(x)$.
[3] $\forall x(A(x) \to B(x)), \forall x(B(x) \to C(x)) \vdash \forall x(A(x) \to C(x))$.
[4] $A \to \forall xB(x) \dashv\vdash \forall x(A \to B(x))$, x not occurring in A.
[5] $A \to \exists xB(x) \dashv\vdash \exists x(A \to B(x))$, x not occurring in A.
[6] $\forall xA(x) \to B \dashv\vdash \exists x(A(x) \to B)$, x not occurring in B.
[7] $\exists xA(x) \to B \dashv\vdash \forall x(A(x) \to B)$, x not occurring in B.

Proof. We choose to prove $\vdash$ of [6].
(1) $\neg\exists x(A(x) \to B) \vdash \forall x\neg(A(x) \to B)$.
(2) $\forall x\neg(A(x) \to B) \vdash \neg(A(u) \to B)$ (use u not occurring in (1)).
(3) $\neg(A(u) \to B) \vdash A(u)$.
(4) $\neg(A(u) \to B) \vdash \neg B$.
(5) $\neg\exists x(A(x) \to B) \vdash A(u)$ (by (1), (2), (3)).
(6) $\neg\exists x(A(x) \to B) \vdash \neg B$ (by (1), (2), (4)).
(7) $\neg\exists x(A(x) \to B) \vdash \forall xA(x)$ (by (5)).
(8) $\forall xA(x) \to B, \neg\exists x(A(x) \to B) \vdash \forall xA(x)$ (by (7)).
(9) $\forall xA(x) \to B, \neg\exists x(A(x) \to B) \vdash \neg B$ (by (6)).
(10) $\forall xA(x) \to B, \neg\exists x(A(x) \to B) \vdash \forall xA(x) \to B$.
(11) $\forall xA(x) \to B, \neg\exists x(A(x) \to B) \vdash B$ (by (10), (8)).
(12) $\forall xA(x) \to B \vdash \exists x(A(x) \to B)$ (by (11), (9)). □

Remarks

In [4] and [5] of Theorem 3.5.4, if x occurs in A, then some quantifier of x would occur in A and accordingly $\forall x(A \to B(x))$ and $\exists x(A \to B(x))$ would not be formulas. It is similar for [6] and [7].

The roman-type capital Latin letter Q (with or without subscripts) will be used for the quantifier symbol $\forall$ or $\exists$.

The proofs of the following Theorems 3.5.5, 3.5.6, and 3.5.7 are left to the reader.

Theorem 3.5.5.

[1] $A \wedge \forall x B(x) \vdash\dashv \forall x(A \wedge B(x))$, x not occurring in A.

[2] $A \wedge \exists x B(x) \vdash\dashv \exists x(A \wedge B(x))$, x not occurring in A.

[3] $\forall x A(x) \wedge \forall x B(x) \vdash\dashv \forall x(A(x) \wedge B(x))$.

[4] $\exists x(A(x) \wedge B(x)) \vdash \exists x A(x) \wedge \exists x B(x)$.

[5] $Q_1 x A(x) \wedge Q_2 y B(y) \vdash\dashv Q_1 x Q_2 y(A(x) \wedge B(y))$, x not occurring in B(y), y not occurring in A(x).

Theorem 3.5.6.

[1] $A \vee \forall x B(x) \vdash\dashv \forall x(A \vee B(x))$, x not occurring in A.

[2] $A \vee \exists x B(x) \vdash\dashv \exists x(A \vee B(x))$, x not occurring in A.

[3] $\forall x A(x) \vee \forall x B(x) \vdash \forall x(A(x) \vee B(x))$.

[4] $\exists x A(x) \vee \exists x B(x) \vdash\dashv \exists x(A(x) \vee B(x))$.

[5] $Q_1 x A(x) \vee Q_2 y B(y) \vdash\dashv Q_1 x Q_2 y(A(x) \vee B(y))$, x not occurring in B(y), y not occurring in A(x).

Theorem 3.5.7.

[1] $\forall x(A(x) \leftrightarrow B(x)) \vdash \forall x A(x) \leftrightarrow \forall x B(x)$.

[2] $\forall x(A(x) \leftrightarrow B(x)) \vdash \exists x A(x) \leftrightarrow \exists x B(x)$.

[3] $\forall x(A(x) \leftrightarrow B(x)), \forall x(B(x) \leftrightarrow C(x)) \vdash \forall x(A(x) \leftrightarrow C(x))$.

[4] $\forall x(A(x) \leftrightarrow B(x)) \vdash \forall x(A(x) \rightarrow B(x))$.

[5] $\forall x(A(x) \leftrightarrow B(x)) \vdash \forall x(B(x) \rightarrow A(x))$.

[6] $\forall x(A(x) \rightarrow B(x)), \forall x(B(x) \rightarrow A(x)) \vdash \forall x(A(x) \leftrightarrow B(x))$.

We write

$$\exists!!x A(x) \quad \text{for} \quad \forall xy(A(x) \wedge A(y) \rightarrow x \approx y).$$

$$\exists!x A(x) \quad \text{for} \quad \exists x[A(x) \wedge \forall y(A(y) \rightarrow x \approx y)].$$

$\exists!!x$ is read as "there exists at most one x such that". It means "there exists at most one individual in the domain such that". It does not mean "there exists".

$\exists!x$ is read as "there exists exactly one x such that". It means "there exists exactly one individual in the domain such that".

Theorem 3.5.8.

[1] $A(t_1), t_1 \approx t_2 \vdash A(t_2)$, where $A(t_2)$ results by replacing some (not necessarily all) occurrences of t_1 in $A(t_1)$ by t_2.

[2] $\emptyset \vdash t \approx t$.

[3] $t_1 \approx t_2 \vdash t_2 \approx t_1$.

[4] $t_1 \approx t_2, t_2 \approx t_3 \vdash t_1 \approx t_3$.

[5] $A(t) \vdash\dashv \forall x(x \approx t \to A(x))$, where $A(x)$ results by replacing some (not necessarily all) occurrences of t in $A(t)$ by x.

[6] $A(t) \vdash\dashv \exists x(x \approx t \wedge A(x))$, where $A(x)$ results as in [5].

[7] $\exists!xA(x) \vdash \exists xA(x), \exists!!xA(x)$.

[8] $\exists xA(x), \exists!!xA(x) \vdash \exists!xA(x)$.

[9] $\exists!xA(x) \vdash\dashv \exists x\forall y(A(y) \leftrightarrow x \approx y)$.

Proof. We choose to prove [6].

Proof of $\vdash$ of [6].

(1) $\emptyset \vdash t \approx t$ (by this theorem [2]).

(2) $A(t) \vdash t \approx t$.

(3) $A(t) \vdash A(t)$.

(4) $A(t) \vdash t \approx t \wedge A(t)$.

(5) $A(t) \vdash \exists x(x \approx t \wedge A(x))$.

Proof of $\dashv$ of [6].

(1) $u \approx t \wedge A(u) \vdash A(u)$ (use u not occurring in $A(t)$).

(2) $u \approx t \wedge A(u) \vdash u \approx t$.

(3) $A(u), u \approx t \vdash A(t)$ (by this theorem [1]).

(4) $u \approx t \wedge A(u) \vdash A(t)$.

(5) $\exists x(x \approx t \wedge A(x)) \vdash A(t)$ (by ($\exists-$), (4)). □

Lemma 3.5.9.

Suppose $A \vdash\dashv A'$, $B \vdash\dashv B'$, and $C(u) \vdash\dashv C'(u)$. Then

[1] $\neg A \vdash\dashv \neg A'$.

[2] $A \wedge B \vdash\dashv A' \wedge B'$.

[3] $A \vee B \vdash\dashv A' \vee B'$.

[4] $A \to B \vdash\dashv A' \to B'$.

[5] $A \leftrightarrow B \vdash\dashv A' \leftrightarrow B'$.

[6] $\forall xC(x) \vdash\dashv \forall xC'(x)$.

[7] $\exists xC(x) \vdash\dashv \exists xC'(x)$.

Proof. [1]–[5] are the same as in Section 2.6. We want to prove [6] and [7].

We have the following:

(1) $C(u) \vdash\dashv C'(u)$ (by supposition).

(2) $\emptyset \vdash C(u) \leftrightarrow C'(u)$.

(3) $\emptyset \vdash \forall x(C(x) \leftrightarrow C'(x))$.

(4) $\emptyset \vdash \forall x C(x) \leftrightarrow \forall x C'(x)$ (by (3), Thm 3.5.7 [1]).

(5) $\emptyset \vdash \exists x C(x) \leftrightarrow \exists x C'(x)$ (by (3), Thm 3.5.7 [2]).

Then [6] and [7] follow from (4) and (5) respectively. □

Suppose B and C are quasi-formulas and B′, C′ are formulas resulting from B, C as in the last section. We write $B \vdash\dashv C$ for $B' \vdash\dashv C'$.

Theorem 3.5.10. (***Replaceability of equivalent formulas***)

Suppose $B \vdash\dashv C$ and A′ results from A by replacing some (not necessarily all) occurrences of B in A by C. Then $A \vdash\dashv A'$.

Proof. By induction on the structure of A, using Lemma 3.5.9. □

Theorem 3.5.11. (***Duality***)

Suppose A is a formula composed of atoms of $\mathcal{L}$, the connectives $\neg$, $\wedge$, $\vee$, and the two quantifiers by the formation rules concerned and A′ is the dual of A. Then $A' \vdash\dashv \neg A$.

Proof. By induction on the structure of A. □

Exercises 3.5.

3.5.1. Prove Theorem 3.5.2 [8].

3.5.2. Prove Theorem 3.5.3 [2].

3.5.3. Prove Theorem 3.5.4 [5], [7].

3.5.4. Prove Theorem 3.5.8 [3], [7]–[9].

3.6. PRENEX NORMAL FORM

Definition 3.6.1. (***Prenex normal form***)
A formula is said to be in *prenex normal form* if it is of the form

$$Q_1x_1 \ldots Q_nx_nB$$

where Q_i ($i = 1, \ldots, n$) is $\forall$ or $\exists$ and B is quantifier-free.
The string $Q_1x_1 \ldots Q_nx_n$ is called the *prefix* and B is called the *matrix*.

A formula with no quantifiers is regarded as a trivial case of a prenex normal form.

Theorem 3.6.2. (***Replaceability of bound variable symbols***)
Suppose A′ results from A by replacing in A some (not necessarily all) occurrences ofQxB(x) by QyB(y). Then $A \models\!\dashv A'$ and $A \vdash\!\dashv A'$.

Proof. First we want to prove

$$(1) \qquad \begin{cases} QxB(x) \models\!\dashv QyB(y). \\ QxB(x) \vdash\!\dashv QyB(y). \end{cases}$$

Obviously x does not occur in QyB(y). y does not occur in QxB(x), otherwise A′ is not well-formed. Therefore x occurs in QxB(x) exactly in the same places as y occurs in QyB(y). Hence (1) holds no matter whether QxB(x) and QyB(y) are formulas or quasi-formulas. By (1) and the replaceability of equivalent formulas, Theorem 3.6.2 is proved. □

Theorem 3.6.3.
Every formula is equivalent to some formula in prenex normal form.

Proof. We have the following:

$\neg\forall xA(x) \models\!\dashv \exists x\neg A(x)$.
$\neg\exists xA(x) \models\!\dashv \forall x\neg A(x)$.
$A \wedge QxB(x) \models\!\dashv Qx(A \wedge B(x))$, x not occurring in A.
$A \vee QxB(x) \models\!\dashv Qx(A \vee B(x))$, x not occurring in A.
$\forall xA(x) \wedge \forall xB(x) \models\!\dashv \forall x(A(x) \wedge B(x))$.
$\exists xA(x) \vee \exists xB(x) \models\!\dashv \exists x(A(x) \vee B(x))$.

$Q_1xA(x) \wedge Q_2yB(y) \models\!\dashv Q_1xQ_2y(A(x) \wedge B(y))$, x not occurring in B(y), y not occurring in A(x).

$Q_1xA(x) \vee Q_2yB(y) \models\!\dashv Q_1xQ_2y(A(x) \vee B(y))$, x not occurring in B(y), y not occurring in A(x).

where the notation $\models\!\dashv$ can be replaced by $\vdash\!\dashv$.

By the theorems of replaceability of equivalent formulas (Theorems 3.4.3 and 3.5.10) and by the equivalent formulas in classical propositional logic, we can replace $\rightarrow$ and $\leftrightarrow$ by $\neg$, $\wedge$, $\vee$ and make $\neg$, $\wedge$, $\vee$ not to occur in the scopes of $\neg$. The above equivalent formulas help to move the quantifiers out of the scopes of the connectives and make them initially placed. Then a prenex normal form is obtained, which is equivalent to the given formula.

Certain bound variable symbols should be replaced when necessary. □

Example

$\neg[\forall x\exists yF(u,x,y) \rightarrow \exists x(\neg\forall yG(y,v) \rightarrow H(x))]$

$\models\!\dashv \neg[\neg\forall x\exists yF(u,x,y) \vee \exists x(\neg\neg\forall yG(y,v) \vee H(x))]$

$\models\!\dashv \neg\neg\forall x\exists yF(u,x,y) \wedge \neg\exists x(\forall yG(y,v) \vee H(x))$

$\models\!\dashv \forall x\exists yF(u,x,y) \wedge \neg\exists x\forall y(G(y,v) \vee H(x))$

$\models\!\dashv \forall x\exists yF(u,x,y) \wedge \forall x\exists y\neg(G(y,v) \vee H(x))$

$\models\!\dashv \forall x[\exists yF(u,x,y) \wedge \exists y(\neg G(y,v) \wedge \neg H(x))]$

$\models\!\dashv \forall x[\exists yF(u,x,y) \wedge \exists z(\neg G(z,v) \wedge \neg H(x))]$

$\models\!\dashv \forall x\exists y\exists z[F(u,x,y) \wedge \neg G(z,v) \wedge \neg H(x)]$.

The matrix of a prenex normal form can be further transformed into a disjunctive or conjunctive normal form.

A prenex normal form equivalent to a formula A is called a prenex normal form of A.

Exercises 3.6.

3.6.1. Transform the following formulas into prenex normal forms:

[1] $(\neg\exists xF(x) \vee \forall yG(y)) \wedge (F(u) \rightarrow \forall zH(z))$

[2] $\exists xF(u,x) \leftrightarrow \forall yG(y)$

[3] $\forall x[F(x) \rightarrow \forall y(F(y) \rightarrow (G(x) \rightarrow G(y)) \vee \forall zF(z))]$

4

AXIOMATIC DEDUCTION SYSTEM

We shall develop in this chapter another type of formal deduction, the axiomatic deduction system, which is essentially based upon formally provable formulas. It is denoted by the notation $\vdash$. $\Sigma \vdash A$ will be demonstrated to be equivalent to $\Sigma \vdash A$.

4.1. AXIOMATIC DEDUCTION SYSTEM

The axiomatic deduction system of formal deduction to be developed in this chapter is essentially based upon formally provable formulas. It consists of some axioms and rules of inference. As will be seen, the axioms are formally provable formulas, and the rules of inference generate formally provable formulas from given ones.

In the following we shall introduce one of such systems for first-order logic with equality.

Axioms.

(Ax1) $A \to (B \to A)$

(Ax2) $(A \to (B \to C)) \to ((A \to B) \to (A \to C))$

(Ax3) $(\neg A \to B) \to ((\neg A \to \neg B) \to A)$

(Ax4) $A \wedge B \to A$

(Ax5) $A \wedge B \to B$

(Ax6) $A \to (B \to A \wedge B)$

(Ax7) $A \to A \vee B$

(Ax8) $A \to B \vee A$

(Ax9) $(A \to C) \to ((B \to C) \to (A \vee B \to C))$

(Ax10) $(A \leftrightarrow B) \to (A \to B)$

(Ax11) $(A \leftrightarrow B) \to (B \to A)$

(Ax12) $(A \to B) \to ((B \to A) \to (A \leftrightarrow B))$

(Ax13) $\forall x A(x) \to A(t)$

(Ax14) $\forall x(A \to B(x)) \to (A \to \forall x B(x))$, x not occurring in A.

(Ax15) $A(t) \to \exists x A(x)$, A(x) resulting from A(t) by replacing some (not necessarily all) occurrences of t in A(t) by x.

(Ax16) $\forall x(A(x) \to B) \to (\exists x A(x) \to B)$, x not occurring in B.

(Ax17) $t_1 \approx t_2 \to (A(t_1) \to A(t_2))$, $A(t_2)$ resulting from $A(t_1)$ by replacing some (not necessarily all) occurrences of t_1 in $A(t_1)$ by t_2.

(Ax18) $u \approx u$

Rules of inference.

(R1) From $A \to B$ and A infer B.

(R2) From A(u) infer $\forall x A(x)$.

(Ax1)–(Ax12) and (R1) belong to propositional logic. The axioms and rules of inference are schemes.

Definition 4.1.1. ($\Sigma \vdash A$)

$\Sigma \vdash A$ (A is *formally deducible* or *provable from* Σ) iff there is some sequence

$$A_1, \ldots, A_n$$

such that each A_k ($k = 1, \ldots, n$) satisfies one of the following:

[1] A_k is an axiom.

[2] $A_k \in \Sigma$.

[3] For some $i, j < k$, $A_i = A_j \to A_k$.

[4] For some $i < k$ and B(u) such that u does not occur in Σ, $A_i = B(u)$ and $A_k = \forall x B(x)$.

and $A_n = A$.

The sequence $A_1, \ldots, A_n$ is called a *formal deduction* (or a *formal proof*) of A *from* Σ.

A is said to be *formally provable* iff $\emptyset \vdash A$ holds.

We have seen in natural deduction (in Chapters 2 and 3) that the rules of formal deduction and the formal proofs resemble those of informal reasoning. But the axioms here do not express the traces of informal reasoning, and hence the formal proofs of formally provable formulas are not natural nor intuitive. This will be illustrated by the following examples.

Example

$\emptyset \vdash A \to A.$

Proof.

(1) $A \to ((A \to A) \to A)$ (by (Ax1))

(2) $[A \to ((A \to A) \to A)] \to [(A \to (A \to A)) \to (A \to A)]$ (by (Ax2))

(3) $(A \to (A \to A)) \to (A \to A)$ (by (R1), (2), (1))

(4) $A \to (A \to A)$ (by (Ax1))

(5) $A \to A$ (by (R1), (3), (4)).

Example

$\emptyset \vdash (B \to C) \to ((A \to B) \to (A \to C)).$

Proof.

(1) $(A \to (B \to C)) \to ((A \to B) \to (A \to C))$ (by (Ax2))

(2) $[(A \to (B \to C)) \to ((A \to B) \to (A \to C))] \to \{(B \to C) \to [(A \to (B \to C)) \to ((A \to B) \to (A \to C))]\}$ (by (Ax1))

(3) $(B \to C) \to [(A \to (B \to C)) \to ((A \to B) \to (A \to C))]$ (by (R1), (2), (1))

(4) $\{(B \to C) \to [(A \to (B \to C)) \to ((A \to B) \to (A \to C))]\} \to \{[(B \to C) \to (A \to (B \to C))] \to [(B \to C) \to ((A \to B) \to (A \to C))]\}$ (by (Ax2))

(5) $[(B \to C) \to (A \to (B \to C))] \to [(B \to C) \to ((A \to B) \to (A \to C))]$ (by (R1), (4), (3))

(6) $(B \to C) \to (A \to (B \to C))$ (by (Ax1))

(7) $(B \to C) \to ((A \to B) \to (A \to C))$ (by (R1), (5), (6)).

In the next section we shall prove the Deduction Theorem:

$$\text{If } \Sigma,\ A \vdash B, \text{ then } \Sigma \vdash A \to B.$$

It is analogous to the rule $(\to +)$ in natural deduction. By Deduction Theorem the formal proofs can be simplified greatly. For instance, the formal proof of

$$\emptyset \vdash (B \to C) \to ((A \to B) \to (A \to C))$$

can be given as follows. First, we prove

$$B \to C, A \to B, A \vdash C$$

by the sequence of formal deduction:

$$B \to C, A \to B, A, B, C.$$

Then we obtain successively by the Deduction Theorem:

$$B \to C, A \to B \vdash A \to C,$$
$$B \to C \vdash (A \to B) \to (A \to C),$$
$$\emptyset \vdash (B \to C) \to ((A \to B) \to (A \to C)).$$

We have mentioned at the beginning of this section that the axioms of the system developed in this chapter are formally provable formulas, and the rules of inference generate formally provable formulas from given ones. After reading Chapter 5, we shall see that the set of formally provable formulas coincides with that of valid ones. Hence the formal deducibility $\vdash$ deals with formally provable formulas (or equivalently valid ones). Formulas in a formal proof of $\emptyset \vdash A$ are formally provable (or equivalently valid). Formulas in a formal proof of $\Sigma \vdash A$ are formally provable (or equivalently valid), if those in Σ are formally provable (or equivalently valid).

We introduce here the axiomatic deduction system of formal deduction because it appeared earlier than natural deduction in the historical development of mathematical logic and is still adopted in the literature.

4.2. RELATION BETWEEN THE TWO DEDUCTION SYSTEMS

In this section we shall demonstrate the equivalence between the two types of formal deducibility:

$$\Sigma \vdash A \quad \text{iff} \quad \Sigma \vdash A.$$

Lemma 4.2.1.
If $\Sigma \models A$, then $\Sigma \vdash A$.

Proof. By $\Sigma \models A$, we suppose

$$A_1, \ldots, A_n \ (= A)$$

is any formal deduction of A from Σ. We can prove $\Sigma \vdash A_k$ ($k = 1, \ldots, n$) by induction. Hence $\Sigma \vdash A$. □

Lemma 4.2.2.
If $\Sigma \vdash A$, then $\Sigma \models A$.

Proof. By induction on the structure of $\Sigma \vdash A$. We have to prove that each of the rules of formal deduction has or preserves the property that if the symbol $\vdash$ in it is replaced by $\models$, the result obtained holds. We choose to prove for the cases of $(\to -)$, $(\to +)$, $(\neg -)$, and $(\exists -)$. The rest are left to the reader.

Case of $(\to -)$. Suppose $\Sigma \models A \to B$ and $\Sigma \models A$. We want to prove $\Sigma \models B$. Let

$$C_1, \ldots, C_k (= A \to B)$$
$$D_1, \ldots, D_l (= A)$$

be, respectively, formal proofs of $A \to B$ and A from Σ (the D's are formulas). Then

$$C_1, \ldots, C_k, D_1, \ldots, D_l, B$$

is a formal proof of B from Σ. Hence $\Sigma \models B$.

Case of $(\to +)$. Suppose $\Sigma, A \models B$, and

$$B_1, \ldots, B_n (= B) \tag{1}$$

is a formal proof of B from Σ and A. We will prove $\Sigma \models A \to B$ by induction. The following subcases need to be considered: B is an axiom or belongs to Σ, B is A, or B is obtained by means of (R1) or (R2) from formulas preceding it in (1).

If B is an axiom or belongs to Σ, the following sequence:

$$B, B \to (A \to B), A \to B$$

forms a formal proof of $A \to B$ from Σ. Hence $\Sigma \vdash A \to B$.

If B is A, the following five formulas:

$$\begin{aligned}
&[A \to ((A \to A) \to A)] \to [(A \to (A \to A)) \to (A \to A)],\\
&A \to ((A \to A) \to A),\\
&(A \to (A \to A)) \to (A \to A),\\
&A \to (A \to A),\\
&A \to A
\end{aligned}$$

form a formal proof of $A \to A$ from Σ. Then $\Sigma \vdash A \to B$.

Suppose B is obtained by means of (R1) from two formulas $C \to B$ and C, which precede B in (1). By the induction hypothesis, we have $\Sigma \vdash A \to (C \to B)$ and $\Sigma \vdash A \to C$. Suppose

$$\begin{aligned}
&D_1, \ldots, D_k \ (= A \to (C \to B))\\
&E_1, \ldots, E_l \ (= A \to C)
\end{aligned}$$

are, respectively, formal proofs of $A \to (C \to B)$ and $A \to C$ from Σ (the E's are formulas). Then the following sequence:

$$\begin{aligned}
&D_1, \ldots, D_k, E_1, \ldots, E_l,\\
&(A \to (C \to B)) \to ((A \to C) \to (A \to B)),\\
&(A \to C) \to (A \to B), A \to B
\end{aligned}$$

forms a formal proof of $A \to B$ from Σ. Hence $\Sigma \vdash A \to B$.

Suppose B is obtained by means of (R2) from a formula C(u) which precedes B in (1), and $B = \forall x C(x)$, then u does not occur in Σ nor in A. By the induction hypothesis, we have $\Sigma \vdash A \to C(u)$. Adding the following three formulas:

$$\begin{aligned}
&\forall x(A \to C(x)),\\
&\forall x(A \to C(x)) \to (A \to \forall x C(x)),\\
&A \to \forall x C(x)
\end{aligned}$$

after a formal proof of $A \to C(u)$ from Σ, we obtain a formal proof of $A \to B$ $(= A \to \forall x C(x))$ from Σ. Hence $\Sigma \vdash A \to B$. Then the proof for the case of $(\to +)$ is completed.

Case of $(\neg-)$. Suppose

$$\Sigma, \neg A \vdash B,$$
$$\Sigma, \neg A \vdash \neg B.$$

We want to prove $\Sigma \vdash A$. By the result proved in the case of $(\to +)$, we have

$$\Sigma \vdash \neg A \to B,$$
$$\Sigma \vdash \neg A \to \neg B.$$

Suppose

$$C_1, \ldots, C_k \; (= \neg A \to B),$$
$$D_1, \ldots, D_l \; (= \neg A \to \neg B)$$

are, respectively, formal proofs of $\neg A \to B$ and $\neg A \to \neg B$ from Σ. Then the sequence

$$C_1, \ldots, C_k, D_1, \ldots, D_l,$$
$$(\neg A \to B) \to ((\neg A \to \neg B) \to A), (\neg A \to \neg B) \to A, A$$

is a formal proof of A from Σ. Hence $\Sigma \vdash A$.

Case of $(\exists-)$. Suppose Σ, $A(u) \vdash B$, u not occurring in Σ nor in B. We want to prove Σ, $\exists x A(x) \vdash B$.

From the result obtained in the case of $(\to +)$, we have $\Sigma \vdash A(u) \to B$. Suppose $C_1, \ldots, C_k$ $(= A(u) \to B)$ is a formal proof of $A(u) \to B$ from Σ. Then the sequence

$$C_1, \ldots, C_k, \forall x(A(x) \to B), \forall x(A(x) \to B) \to (\exists x A(x) \to B),$$
$$\exists x A(x) \to B, \exists x A(x), B$$

is a formal proof of B from Σ and $\exists x A(x)$. Hence Σ, $\exists x A(x) \vdash B$. □

"If Σ, $A \vdash B$, then $\Sigma \vdash A \to B$" is called the *Deduction Theorem*.

Theorem 4.2.3.

$\Sigma \vdash A$ iff $\Sigma \vdash A$. □

5

SOUNDNESS AND COMPLETENESS

We have mentioned in the Introduction that mathematical logic is the study of logical problems and that the (informal) deducibility relations between the premises and conclusions are established by their truth values. Logical consequence, which is defined in terms of valuations, corresponds to (informal) deducibility and involves semantics.

Formal deducibility, which is defined by a finite number of rules of formal deduction, is concerned with the syntactical structures of formulas and involves syntax.

Suppose

1) $$\Sigma \vdash A \Longrightarrow \Sigma \models A$$

for any Σ and A. It is signified by 1) that what formal deducibility expresses about premises and conclusions also holds in informal reasoning, hence formal deducibility does not go beyond the limit of informal reasoning. Then formal deducibility is said to be *sound* for informal reasoning, and 1) is called the *Soundness Theorem.*

Conversely, suppose

2) $$\Sigma \models A \Longrightarrow \Sigma \vdash A$$

for any Σ and A. 2) signifies that what holds in informal reasoning can be expressd in formal deducibility, hence formal deducibility covers informal reasoning. Then formal deducibility is said to be *complete* for informal reasoning, and 2) is called the *Completeness Theorem.*

Soundness and completeness associate the syntactic notion of formal deducibility with the semantic notation of logical consequence, and establish the equivalence between them.

5.1. SATISFIABILITY AND VALIDITY

Satisfiability and validity are important semantical notions which are closely related to soundness and completeness. Definitions of these notions have been formulated in Section 3.3 of Chapter 3.

Theorem 5.1.1.

[1] A is satisfiable iff $\neg$A is invalid.

[2] A is valid iff $\neg$A is unsatisfiable.

Proof. Immediate by the definitions. □

Theorem 5.1.2.

[1] $A(u_1, \ldots, u_n)$ is satisfiable iff $\exists x_1 \ldots x_n A(x_1, \ldots, x_n)$ is satisfiable.

[2] $A(u_1, \ldots, u_n)$ is valid iff $\forall x_1 \ldots x_n A(x_1, \ldots, x_n)$ is valid.

Proof. For simplicity we will prove without loss of generality, instead of [1] and [2], the following:

(1) A(u) is satisfiable iff $\exists$xA(x) is satisfiable.

(2) A(u) is valid iff $\forall$xA(x) is valid.

We first prove (1). Suppose A(u) is satisfiable, that is, there is some valuation v over domain D such that $A(u)^v = 1$. Obviously, $u^v \in D$, $v(u/u^v)$ is identical with v, and $A(u)^{v(u/u^v)} = 1$. Then we have $\exists xA(x)^v = 1$ and $\exists$xA(x) is satisfiable.

Suppose $\exists$xA(x) is satisfied by some valuation v over domain D. Then there is some $\alpha \in D$ such that $A(u)^{v(u/\alpha)} = 1$. Hence A(u) is satisfiable, and (1) is proved.

(2) can be proved in an analogous way. It can also be proved by Theorem 5.1.1 and (1) as follows:

$$\begin{aligned}
&\mathrm{A(u)} \text{ is valid}\\
&\Longleftrightarrow \neg\mathrm{A(u)} \text{ is unsatisfiable}\\
&\Longleftrightarrow \exists \mathrm{x}\neg\mathrm{A(x)} \text{ is unsatisfiable}\\
&\Longleftrightarrow \neg\forall \mathrm{x}\mathrm{A(x)} \text{ is unsatisfiable}\\
&\Longleftrightarrow \forall \mathrm{x}\mathrm{A(x)} \text{ is valid.} \quad \square
\end{aligned}$$

Because any formula A is equivalent to its prenex normal form, we have

Theorem 5.1.3.
[1] A is satisfiable iff the prenex normal form of A is satisfiable.
[2] A is valid iff the prenex normal form of A is valid. □

Definition 5.1.4. (***Satisfiability, validity in a domain***)
Suppose $\Sigma \subseteq Form(\mathcal{L})$, $\mathrm{A} \in Form(\mathcal{L})$, and D is a domain.
[1] Σ is *satisfiable in D* iff there is some valuation v over D such that $\Sigma^v = 1$.
[2] A is *valid in D* iff for every valuation v over D, $\mathrm{A}^v = 1$.

Obviously we have the following corollaries:

$$\begin{aligned}
&\Sigma \text{ is satisfiable in } D \Longrightarrow \Sigma \text{ is satisfiable.}\\
&\mathrm{A} \text{ is valid} \Longrightarrow \mathrm{A} \text{ is valid in } D.\\
&\mathrm{A} \text{ is satisfiable in } D \Longleftrightarrow \neg\mathrm{A} \text{ is invalid in } D.\\
&\mathrm{A} \text{ is valid in } D \Longleftrightarrow \neg\mathrm{A} \text{ is unsatisfiable in } D.\\
&\mathrm{A}(\mathrm{u}_1,\ldots,\mathrm{u}_n) \text{ is satisfiable in } D\\
&\quad \Longleftrightarrow \exists \mathrm{x}_1\ldots\mathrm{x}_n\mathrm{A}(\mathrm{x}_1,\ldots,\mathrm{x}_n) \text{ is satisfiable in } D.\\
&\mathrm{A}(\mathrm{u}_1,\ldots,\mathrm{u}_n) \text{ is valid in } D\\
&\quad \Longleftrightarrow \forall \mathrm{x}_1\ldots\mathrm{x}_n\mathrm{A}(\mathrm{x}_1,\ldots,\mathrm{x}_n) \text{ is valid in } D.\\
&\mathrm{A} \text{ is satisfiable in } D\\
&\quad \Longleftrightarrow \text{The prenex normal form of A is satisfiable in } D.\\
&\mathrm{A} \text{ is valid in } D\\
&\quad \Longleftrightarrow \text{The prenex normal form of A is valid in } D.
\end{aligned}$$

Now we want to prove Theorem 5.1.7. For this purpose we will need to have the following preparations, including two lemmas.

Suppose D and D_1 are two domains such that $|D| \leq |D_1|$. Suppose $D' \subseteq D_1$ such that D and D' are in one-one correspondence and $\alpha \in D$ corresponds to $\alpha' \in D'$.

Suppose γ is an arbitrary element of D. For every $\beta \in D_1$, a unique $\beta^* \in D$ is assigned as follows:

1) $$\beta^* = \begin{cases} \alpha & \text{if } \beta = \alpha' \in D', \\ \gamma & \text{if } \beta \notin D'. \end{cases}$$

Suppose A is satisfiable in D, that is, there is some valuation v over D such that $\mathrm{A}^v = 1$. Construct a valuation v_1 over D_1 satisfying the following conditions 2)–5):

2) $\mathrm{a}^{v_1} = (\mathrm{a}^v)'$.
3) $\mathrm{u}^{v_1} = (\mathrm{u}^v)'$.
4) For any $\beta_1, \ldots, \beta_n \in D_1$, $\langle \beta_1, \ldots, \beta_n \rangle \in \mathrm{F}^{v_1}$ iff $\langle \beta_1^*, \ldots, \beta_n^* \rangle \in \mathrm{F}^v$.
5) For any $\beta_1, \ldots, \beta_n \in D_1$, $\mathrm{f}^{v_1}(\beta_1, \ldots, \beta_n) = \mathrm{f}^v(\beta_1^*, \ldots, \beta_n^*)'$.

Lemma 5.1.5.

Suppose

[1] The individual symbols and free variable symbols occurring in a term t are included in $\mathrm{a}_1, \ldots, \mathrm{a}_k$, $\mathrm{u}_1, \ldots, \mathrm{u}_l$.

[2] v_1^* is a valuation over D such that $\mathrm{a}_i^{v_1^*} = (\mathrm{a}_i^{v_1})^* (i = 1, \ldots, k)$, $\mathrm{u}_j^{v_1^*} = (\mathrm{u}_j^{v_1})^* (j = 1, \ldots, l)$, and v_1^* agrees with v on all the function symbols occurring in t.

Then,

[3] $(\mathrm{t}^{v_1})^* = \mathrm{t}^{v_1^*}$.

Proof. By induction on the structure of t. For the simplicity of description we may consider without loss of generality only one individual symbol a and one free variable symbol u occurring in t.

Basis. $\mathrm{t} = \mathrm{a}$ or $\mathrm{t} = \mathrm{u}$. If $\mathrm{t} = \mathrm{a}$, then by [2] we have

$$(\mathrm{t}^{v_1})^* = (\mathrm{a}^{v_1})^* = \mathrm{a}^{v_1^*} = \mathrm{t}^{v_1^*}.$$

Similarly for the case of $\mathrm{t} = \mathrm{u}$. Hence [3] holds.

Induction step. $t = f(t_1)$. (For simplicity we regard f as unary.) Then,

$$
\begin{aligned}
&(f(t_1)^{v_1})^* \\
&= (f^{v_1}(t_1^{v_1}))^* \\
&= (f^{v}((t_1^{v_1})^*))'^* \quad \text{(by 5))} \\
&= f^{v}((t_1^{v_1})^*) \quad \text{(by 1))} \\
&= f^{v}(t_1^{v_1^*}) \quad \text{(by ind hyp)} \\
&= f^{v_1^*}(t_1^{v_1^*}) \quad \text{(by [2])} \\
&= f(t_1)^{v_1^*}.
\end{aligned}
$$

Hence [3] holds. □

Lemma 5.1.6.

Suppose

[1] The individual symbols and free variable symbols occurring in a formula A without the equality symbol are included in $a_1, \ldots, a_k$, $u_1, \ldots, u_l$.

[2] Same as in Lemma 5.1.5 except that v_1^* agrees with v on all the function and relation symbols occurring in A.

Then,

[3] $A^{v_1} = A^{v_1^*}$.

Proof. By induction on the structure of A.

Basis. A is an atom $F(t)$. (For simplicity we regard F as unary.) Then,

$$
\begin{aligned}
F(t)^{v_1} = 1 &\iff t^{v_1} \in F^{v_1} \\
&\iff (t^{v_1})^* \in F^{v} \quad \text{(by 4))} \\
&\iff t^{v_1^*} \in F^{v} \quad \text{(by Lem 5.1.5)} \\
&\iff t^{v_1^*} \in F^{v_1^*} \quad \text{(by [2])} \\
&\iff F(t)^{v_1^*} = 1.
\end{aligned}
$$

Hence $F(t)^{v_1} = F(t)^{v_1^*}$ and [3] holds.

Induction step. We distinguish seven cases: $A = \neg B$, $B \wedge C$, $B \vee C$, $B \to C$, $B \leftrightarrow C$, $\forall x B(x)$, or $\exists x B(x)$. We will prove [3] for the cases of $\neg B$, $B \vee C$, and $\exists x B(x)$ and leave the rest to the reader.

Case of $A = \neg B$.

$$\begin{aligned} &(\neg B)^{v_1} = 1 \\ \iff &B^{v_1} = 0 \\ \iff &B^{v_1^*} = 0 \quad \text{(by ind hyp)} \\ \iff &(\neg B)^{v_1^*} = 1. \end{aligned}$$

Case of $A = B \vee C$.

$$\begin{aligned} &(B \vee C)^{v_1} = 1 \\ \iff &B^{v_1} = 1 \quad \text{or} \quad C^{v_1} = 1 \\ \iff &B^{v_1^*} = 1 \quad \text{or} \quad C^{v_1^*} = 1 \\ \iff &(B \vee C)^{v_1^*} = 1. \end{aligned}$$

Case of $A = \exists x B(x)$.

We are to prove $\exists x B(x)^{v_1} = 1$ iff $\exists x B(x)^{v_1^*} = 1$.

Choose any v not occurring in $\exists x B(x)$ and form $B(v)$ from $B(x)$. Then, in addition to the non-logical symbols and free variable symbols occurring in $\exists x B(x)$, $B(v)$ contains one more free variable symbol v.

Suppose $\exists x B(x)^{v_1} = 1$, that is,

(1) There is some $\beta \in D_1$ such that $B(v)^{v_1(v/\beta)} = 1$.

where $v_1(v/\beta)$ is a valuation over D_1, which is exactly the same as v_1 except that $v^{v_1(v/\beta)} = \beta$.

From $\beta \in D_1$ we obtain $\beta^* \in D$. Construct a valuation $v_1^*(v/\beta^*)$ over D,which is exactly the same as v_1^* except that $v^{v_1^*(v/\beta^*)} = \beta^*$. $v_1^*(v/\beta^*)$ and $v_1(v/\beta)$ are respectively the result of extending v_1^* and v_1 from the valuation of non-logical symbols and free variable symbols in $\exists x B(x)$ to v (v occurring in $B(v)$, but not in $\exists x B(x)$). For v, we have

$$v^{v_1^*(v/\beta^*)} = \beta^* = (v^{v_1(v/\beta)})^* .$$

Hence the relation between $v_1^*(v/\beta^*)$ and $v_1(v/\beta)$ is the same as that between v_1^* and v_1. Then, by induction hypothesis, we have

$$B(v)^{v_1^*(v/\beta^*)} = B(v)^{v_1(v/\beta)}. \tag{2}$$

From (1) and (2) we obtain $B(v)^{v_1^*(v/\beta^*)} = 1$, where $\beta^* \in D$. Hence $\exists x B(x)^{v_1^*} = 1$.

To prove the converse, suppose $\exists x B(x)^{v_1^*} = 1$, that is,

(3) There is some $\alpha \in D$ such that $B(v)^{v_1^*(v/\alpha)} = 1$.

From $\alpha \in D$ we obtain $\alpha' \in D_1$. Construct a valuation $v_1(\mathrm{v}/\alpha')$ over domain D_1. As in the previous case, $v_1^*(\mathrm{v}/\alpha)$ and $v_1(\mathrm{v}/\alpha')$ are respectively the result of extending v_1^* and v_1 to v. For v we have

$$\mathrm{v}^{v_1^*(\mathrm{v}/\alpha)} = \alpha = \alpha'^* = (\mathrm{v}^{v_1(\mathrm{v}/\alpha')})^* .$$

Hence the relation between $v_1^*(\mathrm{v}/\alpha)$ and $v_1(\mathrm{v}/\alpha')$ is the same as that between v_1^* and v_1. Then, by induction hypothesis, we have

$$\text{(4)} \qquad \mathrm{B}(\mathrm{v})^{v_1^*(\mathrm{v}/\alpha)} = \mathrm{B}(\mathrm{v})^{v_1(\mathrm{v}/\alpha')} .$$

From (3) and (4) we obtain $\mathrm{B}(\mathrm{v})^{v_1(\mathrm{v}/\alpha')} = 1$, where $\alpha' \in D_1$. Hence $\exists \mathrm{x}\mathrm{B}(\mathrm{x})^{v_1} = 1$. [3] is proved in the induction step. □

Theorem 5.1.7.
Suppose A contains no equality symbol and $|D| \leq |D_1|$.
[1] If A is satisfiable in D, then A is satisfiable in D_1.
[2] If A is valid in D_1, then A is valid in D.

Proof. Suppose A is satisfiable in D, that is, there is some valuation v over D such that

$$\text{(1)} \qquad \mathrm{A}^v = 1.$$

By the conventions, notations, and results stated in Lemmas 5.1.5 and 5.1.6, we have

(2) $\mathrm{A}^{v_1} = \mathrm{A}^{v_1^*}$.
(3) $\mathrm{a}_i^{v_1^*} = (\mathrm{a}_i^{v_1})^* = (\mathrm{a}_i^v)'^* = \mathrm{a}_i^v \quad (i = 1, \ldots, k)$.
(4) $\mathrm{u}_j^{v_1^*} = (\mathrm{u}_j^{v_1})^* = (\mathrm{u}_j^v)'^* = \mathrm{u}_j^v \quad (j = 1, \ldots, l)$.

Since v_1^* agrees with v on all the function and relation symbols occurring in A, we have by (3), (4) and Theorem 3.3.6,

$$\text{(5)} \qquad \mathrm{A}^{v_1^*} = \mathrm{A}^v.$$

Then $\mathrm{A}^{v_1} = 1$ by (2), (5), and (1). Hence A is satisfiable in D_1, and [1] is proved.

[2] follows immediately from [1]. □

Note that the formula in Theorem 5.1.7 contains no equality symbol and a set Σ can be used instead of A in [1]. As counterexamples: $\forall xy(x \approx y)$ is satisfiable in a domain with one individual, but unsatisfiable in domains with more individuals; $\exists xy\neg(x \approx y)$ is valid in domains with two or more individuals, but invalid in a domain with one individual.

Exercises 5.1.

5.1.1. Construct a sentence such that

[1] It is valid in domains with one individual but invalid in larger ones.

[2] It is valid in domains with one or two individuals but invalid in larger ones.

[3] It is valid in domains with one or two or three individuals but invalid in larger ones.

5.1.2. Construct a sentence satisfiable in a domain D iff

[1] D has one individual.

[2] D has two individuals.

[3] D has three individuals.

[4] D has at least three individuals.

[5] D has at most three individuals.

5.1.3. The sentence

$$\forall x\exists y F(x,y) \land \forall x\neg F(x,x) \\ \land \forall xyz(F(x,y) \land F(y,z) \to F(x,z))$$

is satisfiable in infinite domains but unsatisfiable in finite ones.

5.1.4. The sentences

[1] $\exists x\forall y\exists z[(F(y,z) \to F(x,z)) \to (F(x,x) \to F(y,x))]$

[2] $\forall xF(x,x) \land \forall xyz(F(x,z) \to F(x,y) \lor F(y,z)) \to \exists x\forall yF(x,y)$

are valid in finite domains but invalid in infinite ones.

5.1.5. The sentence

$$\exists x\forall y[F(x,y) \land \neg F(y,x) \to (F(x,x) \leftrightarrow F(y,y))]$$

is valid in domains with no more than three individuals but invalid in domains with four individuals.

5.2. SOUNDNESS

Theorem 5.2.1. (***Soundness***)

[1] If $\Sigma \vdash A$, then $\Sigma \models A$.

[2] If $\emptyset \vdash A$, then $\emptyset \models A$.

(That is, every formally provable formula is valid.)

Proof. [1] will be proved by induction on the structure of $\Sigma \vdash A$. That is, we are to prove that each of the seventeen rules of formal deduction of first-order logic has or preserves the property: the statement obtained by replacing $\vdash$ by $\models$ in each rule holds. We will prove the cases of (Ref), (+), ($\neg-$), ($\vee-$), and ($\exists-$). The rest are left to the reader.

Case of (Ref). $A \models A$ is obvious.

Case of (+). It is also obvious that if $\Sigma \models A$, then $\Sigma, \Sigma' \models A$.

Case of ($\neg-$). We shall prove:

$$\begin{aligned} \text{If}\quad & \Sigma, \neg A \models B, \\ & \Sigma, \neg A \models \neg B, \\ \text{then}\quad & \Sigma \models A. \end{aligned}$$

Suppose $\Sigma \not\models A$, that is, there is some valuation v such that $\Sigma^v = 1$ and $A^v = 0$. Then $(\neg A)^v = 1$. Since Σ, $\neg A \models B$ and Σ, $\neg A \models \neg B$, we have $B^v = 1$ and $(\neg B)^v = 1$, which is a contradiction. Hence $\Sigma \models A$.

Case of ($\vee-$). We shall prove:

$$\begin{aligned} \text{If}\quad & \Sigma, A \models C, \\ & \Sigma, B \models C, \\ \text{then}\quad & \Sigma, A \vee B \models C. \end{aligned}$$

Suppose v is an arbitrary valuation such that $\Sigma^v = 1$ and $(A \vee B)^v = 1$. Then $A^v = 1$ or $B^v = 1$. If $A^v = 1$, then by Σ, $A \models C$ we have $C^v = 1$. If $B^v = 1$, then by Σ, $B \models C$ we have $C^v = 1$. Hence $C^v = 1$ and accordingly Σ, $A \vee B \models C$.

Case of ($\exists-$). We shall prove:

$$\begin{aligned} \text{If}\quad & \Sigma, A(u) \models B, u \text{ not occurring in } \Sigma \text{ or } B, \\ \text{then}\quad & \Sigma, \exists x A(x) \models B. \end{aligned}$$

Suppose v is an arbitrary valuation over domain D such that $\Sigma^v = 1$ and $\exists x A(x)^v = 1$. Then there is some $\alpha \in D$ such that $A(u)^{v(u/\alpha)} = 1$. Since u does not occur in Σ, we have $\Sigma^{v(u/\alpha)} = \Sigma^v = 1$. By $\Sigma, A(u) \models B$, we have $B^{v(u/\alpha)} = 1$. Since u does not occur in B, we have $B^v = B^{v(u/\alpha)} = 1$. Hence $\Sigma, \exists x A(x) \models B$.

Thus, [1] is proved. [2] is a special case of [1]. □

Remarks

In the case of $(\neg -)$ in the above proof we have proved

$$(1) \qquad \begin{cases} \text{If} \quad \Sigma, \neg A \models B, \\ \qquad \Sigma, \neg A \models \neg B, \\ \text{then} \quad \Sigma \models A. \end{cases}$$

which expresses the method of indirect proof. In proving (1) we have used the method of indirect proof. It seems that the method of indirect proof is proved by itself. In fact (1) is the method of indirect proof expressed in the object language, while what we have used in proving (1) is the method of indirect proof which takes place in the metalanguage.

It is similar for the case of $(\vee -)$.

Definition 5.2.2. (***Consistency***)

$\Sigma \subseteq Form(\mathcal{L})$ is *consistent* iff there is no $A \in Form(\mathcal{L})$ such that $\Sigma \vdash A$ and $\Sigma \vdash \neg A$.

Note that consistency is a syntactical notion.

Theorem 5.2.3. (***Soundness***)

If Σ is satisfiable, then Σ is consistent.

Proof. Left as an exercise. □

Theorem 5.2.1 is the Soundness Theorem formulated in terms of logical consequence (a semantical notion) and formal deducibility or provability (a syntactical notion). Theorem 5.2.3 is an equivalent version of the Soundness Theorem formulated in terms of satisfiability (a semantical notion) and consistency (a syntactical notion).

Theorem 5.2.3 illustrates why in mathematical practice the consistency of a theory is established by exhibiting a model. (A model of a theory, denoted by Σ, is a valuation satisfying it.)

Exercises 5.2.

5.2.1. Suppose Σ is finite. Prove "$\Sigma \vdash A \Longrightarrow \Sigma \models A$" from "Every formally provable formula is valid".

5.2.2. Σ is consistent iff there is some A such that $\Sigma \nvdash A$.

5.2.3. Theorem 5.2.1 [1] is equivalent to Theorem 5.2.3. Theorem 5.2.1 [2] is equivalent to "Every satisfiable formula is consistent".

5.2.4. Which of the following sets are consistent?
[1] $\{A \wedge (B \to C), A \to (B \wedge C), \neg B \leftrightarrow C\}$
[2] $\{A \to B, B \to C, C \to C_1, C_1 \to \neg A\}$

5.2.5. Σ is said to be independent iff for each $A \in \Sigma$, $\Sigma - \{A\} \nvdash A$. Prove in propositional logic
[1] Each finite Σ has an independent $\Delta \subseteq \Sigma$ such that $\Delta \vdash A$ for all $A \in \Sigma$.
[2] Let $\Sigma = \{A_1, A_2, A_3, \ldots\}$. Find an equivalent set $\Delta = \{B_1, B_2, B_3, \ldots\}$ (that is, for all i, $\Sigma \vdash B_i$ and $\Delta \vdash A_i$) such that $B_{n+1} \vdash B_n$ but $B_n \nvdash B_{n+1}$ $(n \geq 1)$.

5.3. COMPLETENESS OF PROPOSITIONAL LOGIC

The proof of completeness of propositional logic, based on the truth table method, was first made by Post in 1921. Since then a number of different proofs have been published. The proof mentioned here is an adaptation to propositional logic of the method used by Henkin in proving the completeness of first-order logic.

We begin with the notion of a maximal consistent set (of formulas) and some of its properties. "Consistency" and "consistent" will sometimes be abbreviated as "consis", and "maximal consistency" and "maximal consitent" abbreviated as "max consis".

Definition 5.3.1. (***Maximal consistency***)
$\Sigma \subseteq Form(\mathcal{L}^p)$ is *maximal consistent* iff
[1] Σ is consistent.
[2] For any $A \in Form(\mathcal{L}^p)$ such that $A \notin \Sigma$, $\Sigma \cup \{A\}$ is inconsistent.

[2] in Definition 5.3.1 is equivalent to "there is no consistent set which contains Σ as a proper subset".

Lemma 5.3.2.
Suppose Σ is maximal consistent. Then $A \in \Sigma$ iff $\Sigma \vdash A$.

Proof. If $A \in \Sigma$, then $\Sigma \vdash A$ by (ϵ). For the converse, suppose $\Sigma \vdash A$ and $A \notin \Sigma$. Since Σ is maximal consistent, $\Sigma \cup \{A\}$ is inconsistent by Definition 5.3.1. Then $\Sigma \vdash \neg A$ and Σ is inconsistent, contradicting the maximal consistency of Σ. Hence $A \in \Sigma$. □

Σ is said to be *closed under formal deducibility* if $\Sigma \vdash A$ implies $A \in \Sigma$.

Lemma 5.3.3.
Suppose Σ is maximal consistent. Then
[1] $\neg A \in \Sigma$ iff $A \notin \Sigma$.
[2] $A \wedge B \in \Sigma$ iff $A \in \Sigma$ and $B \in \Sigma$.
[3] $A \vee B \in \Sigma$ iff $A \in \Sigma$ or $B \in \Sigma$.
[4] $A \rightarrow B \in \Sigma$ iff $A \in \Sigma$ implies $B \in \Sigma$.
[5] $A \leftrightarrow B \in \Sigma$ iff $A \in \Sigma$ iff $B \in \Sigma$.

Proof. We will prove [1] and [3].

Proof of [1]. Suppose $\neg A \in \Sigma$ and $A \in \Sigma$. By (ϵ) we have $\Sigma \vdash A$ and $\Sigma \vdash \neg A$, that is, Σ is inconsistent, contradicting the supposition that Σ is maximal consistent. Hence $\neg A \in \Sigma \Longrightarrow A \notin \Sigma$.

For the converse, suppose $A \notin \Sigma$ and $\neg A \notin \Sigma$. Then we have

$$\begin{array}{ll} \Sigma \cup \{A\} & \text{is inconsistent.} \\ \Sigma \vdash \neg A & \text{(by } (\neg +)). \\ \neg A \in \Sigma & \text{(by Lem 5.3.2).} \end{array}$$

which contradict $\neg A \notin \Sigma$. Hence $A \notin \Sigma \Longrightarrow \neg A \in \Sigma$.

Proof of [3]. By Lemma 5.3.2 and $(\vee +)$ we have

$$A \in \Sigma \Longrightarrow \Sigma \vdash A \Longrightarrow \Sigma \vdash A \vee B \Longrightarrow A \vee B \in \Sigma.$$
$$B \in \Sigma \Longrightarrow \Sigma \vdash B \Longrightarrow \Sigma \vdash A \vee B \Longrightarrow A \vee B \in \Sigma.$$

Hence "$A \in \Sigma$ or $B \in \Sigma$" implies $A \vee B \in \Sigma$.

For the converse, suppose $A \vee B \in \Sigma$ but not "$A \in \Sigma$ or $B \in \Sigma$". Then we have

$$A, B \notin \Sigma.$$
$$\neg A, \neg B \in \Sigma \quad \text{(by Lem 5.3.3 [1])}.$$
$$\neg A \wedge \neg B \in \Sigma \quad \text{(by Lem 5.3.3 [2])}.$$
$$\Sigma \vdash \neg A \wedge \neg B.$$
$$\Sigma \vdash \neg(A \vee B).$$
$$\Sigma \vdash A \vee B \quad \text{(by } A \vee B \in \Sigma).$$

Thus Σ is inconsistent, contradicting the maximal consistency of Σ. Hence $A \vee B \in \Sigma$ implies "$A \in \Sigma$ or $B \in \Sigma$". □

Lemma 5.3.4.
Suppose Σ is maximal consistent. Then $\Sigma \vdash \neg A$ iff $\Sigma \nvdash A$. □

Lemma 5.3.5. (***Lindenbaum***)
Any consistent set of formulas can be extended to some maximal consistent set.

Proof. Suppose Σ is consistent, and

$$A_1, A_2, A_3, \ldots \tag{1}$$

is an arbitrary enumeration of $Form(\mathcal{L}^p)$. Construct an infinite sequence of sets $\Sigma_n \subseteq Form(\mathcal{L}^p)$ as follows ($n \geq 0$):

$$\begin{cases} \Sigma_0 = \Sigma \\ \Sigma_{n+1} = \begin{cases} \Sigma_n \cup \{A_{n+1}\} & \text{if } \Sigma_n \cup \{A_{n+1}\} \text{ is consistent,} \\ \Sigma_n & \text{otherwise.} \end{cases} \end{cases} \tag{2}$$

Then we have

(3) $\Sigma_n \subseteq \Sigma_{n+1}$.
(4) Σ_n is consistent.

where (3) is obvious, and (4) can be proved by induction on n.

Suppose $\Sigma^* = \bigcup_{n \in N} \Sigma_n$. We want to prove that Σ^* is the maximal consistent set required in this lemma.

We first prove that Σ^* is consistent. Suppose Σ^* is inconsistent. Then there is some finite subset $\{B_1, \ldots, B_k\}$ of Σ^* which is inconsistent. Suppose $B_1 \in \Sigma_{i_1}, \ldots, B_k \in \Sigma_{i_k}$ and $i = max(i_1, \ldots, i_k)$. By (3) we have $\{B_1, \ldots, B_k\} \subseteq \Sigma_i$. Then Σ_i is inconsistent, contradicting (4). Hence Σ^* is consistent.

Suppose $B \notin \Sigma^*$, that is, $B \notin \Sigma_n$ $(n = 0, 1, 2, \ldots)$. B is a formula, say A_{m+1}, in (1). By (2) the set $\Sigma_m \cup \{A_{m+1}\}$ (that is, $\Sigma_m \cup \{B\}$) is inconsistent. Then $\Sigma^* \cup \{B\}$ is inconsistent because $\Sigma_m \subseteq \Sigma^*$. Therefore Σ^* is maximal consistent. □

Note that since the enumeration (1) in the above proof is arbitrary, the Σ^* constructed is not unique.

Remarks

$\mathcal{L}^P$ (and $\mathcal{L}$ as well) is countably infinite. A formal language may be non-countable such that the set of its formulas is non-countable. In such case its formulas can be arranged as a well-ordered set (suppose its order-type is α)

$$\{A_0, A_1, A_2, \ldots, A_\beta, \ldots\} \quad (\beta < \alpha).$$

Beginning with a given consistent set Σ of formulas, we define increasing consistent sets

$$\Sigma_0 \subseteq \Sigma_1 \subseteq \Sigma_2 \subseteq \ldots \subseteq \Sigma_\beta \subseteq \ldots (\beta < \alpha)$$

as follows.

Set $\Sigma_0 = \Sigma$.

For any $\beta < \alpha$, suppose for every ordinal $\delta < \beta$, Σ_δ is defined and consistent. Then Σ_β is defined as follows.

1) Suppose β is a successor ordinal $\gamma + 1$. Set

$$\Sigma_\beta = \begin{cases} \Sigma_\gamma \cup \{A_\gamma\} & \text{if } \Sigma_\gamma \cup \{A_\gamma\} \text{ is consistent,} \\ \Sigma_\gamma & \text{otherwise.} \end{cases}$$

2) Suppose β is a limit ordinal. Set

$$\Sigma_\beta = \bigcup_{\delta < \beta} \Sigma_\delta .$$

We define

$$\Sigma^* = \bigcup_{\beta<\alpha} \Sigma_\beta .$$

Σ^* can be proved to be maximal consistent.

Non-countability and well-ordering are concepts not contained in this book. Readers may refer to books in set theory about them.

Lemma 5.3.6.

Suppose $\Sigma^* \subseteq Form(\mathcal{L}^p)$ is maximal consistent and t is a truth valuation such that for every atom p, $p^t = 1$ iff $p \in \Sigma^*$. Then for every $A \in Form(\mathcal{L}^p)$, $A^t = 1$ iff $A \in \Sigma^*$.

Proof. By induction on the structure of A.

Basis. A is an atom. The lemma holds by supposition.

Induction step. We distinguish five cases: $A = \neg B$, $B \wedge C$, $B \vee C$, $B \to C$ or $B \leftrightarrow C$.

Case of $A = \neg B$.

$$\begin{aligned} & \neg B \in \Sigma^* \\ \Longleftrightarrow\ & B \notin \Sigma^* \quad \text{(by Lem 5.3.3 [1])} \\ \Longleftrightarrow\ & B^t = 0 \quad \text{(by ind hyp)} \\ \Longleftrightarrow\ & (\neg B)^t = 1. \end{aligned}$$

Case of $A = B \vee C$.

$$\begin{aligned} & B \vee C \in \Sigma^* \\ \Longleftrightarrow\ & B \in \Sigma^* \quad \text{or} \quad C \in \Sigma^* \quad \text{(by Lem 5.3.3 [3])} \\ \Longleftrightarrow\ & B^t = 1 \quad \text{or} \quad C^t = 1 \quad \text{(by ind hyp)} \\ \Longleftrightarrow\ & (B \vee C)^t = 1. \end{aligned}$$

The other cases are left to the reader. Thus, the induction step is proved. □

Theorem 5.3.7. (***Completeness***)

If Σ is consistent, then Σ is satisfiable.

Proof. Suppose Σ is consistent and $A \in \Sigma$. Extend Σ to some maximal consistent set Σ^*. Then $A \in \Sigma^*$. By Lemma 5.3.6, $A^t = 1$. Hence Σ is satisfied by t. □

Theorem 5.3.8. (***Completeness***)
[1] If $\Sigma \models$ A, then $\Sigma \vdash$ A.
[2] If $\emptyset \models$ A, then $\emptyset \vdash$ A.
(That is, every tautology is formally provable.)

Proof. Suppose $\Sigma \models$ A. Then $\Sigma \cup \{\neg A\}$ is unsatisfiable. By Theorem 5.3.7, $\Sigma \cup \{\neg A\}$ is inconsistent. Hence $\Sigma \vdash$ A holds and [1] is proved.
[2] is a special case of [1]. □

Theorem 5.3.7 and 5.3.8 are equivalent versions of the Completeness Theorem. One is formulated in terms of satisfiability and consistency and the other in terms of logical consequence and formal deducibility or provability.

Exercises 5.3.

5.3.1. Prove that [1] of Theorem 5.3.8 is equivalent to Theorem 5.3.7, and [2] is equivalent to "Every consistent formula is satisfiable".

5.3.2. Prove by means of normal form that every tautology is formally provable.

5.3.3. Suppose A contains distinct atoms $p_1, \ldots, p_n$ and t is a truth valuation. For $i = 1, \ldots, n$, let

$$A_i = \begin{cases} p_i & \text{if } p_i^t = 1, \\ \neg p_i & \text{otherwise.} \end{cases}$$

Prove
[1] $A^t = 1 \Longrightarrow A_1, \ldots, A_n \vdash A$.
[2] $A^t = 0 \Longrightarrow A_1, \ldots, A_n \vdash \neg A$.

5.3.4. Prove by Exercise 5.3.3 that every tautology is formally provable.

5.3.5. Suppose Σ is closed under formal deducibility. Prove that Σ is maximal consistent iff for any A, Σ contains exactly one of A and $\neg$A.

5.3.6. Suppose $\Sigma \subseteq Form(\mathcal{L}^p)$ is closed under formal deducibility. Prove that Σ is maximal consistent iff there is a unique truth valuation t such that $\Sigma^t = 1$.

5.4. COMPLETENESS OF FIRST-ORDER LOGIC

The Completeness Theorem is the most important and profound theorem of first-order logic. It was first proved by Gödel [1930], and hence is called Gödel's Completeness Theorem. The proof stated here is due to Henkin [1949].

We will extend the first-order language $\mathcal{L}$ without equality to $\mathcal{L}^\circ$ by adding to $\mathcal{L}$ an infinite sequence of new free variable symbols. The roman-type small Latin letter

$$\mathrm{d}$$

(with or without subscripts) is used for any one of such new symbols. Then $Term(\mathcal{L})$, $Atom(\mathcal{L})$, and $Form(\mathcal{L})$ are proper subsets of $Term(\mathcal{L}^\circ)$, $Atom(\mathcal{L}^\circ)$, and $Form(\mathcal{L}^\circ)$, which are sets of terms, atoms, and formulas of $\mathcal{L}^\circ$ respectively.

Equality will not be treated for the time being.

Definition 5.4.1. (***E-property***)

Suppose $\Sigma \subseteq Form(\mathcal{L}^\circ)$. Σ is said to have the *existence property* (abbreviated as *E-property*) iff for every existential formula $\exists \mathrm{x} \mathrm{A}(\mathrm{x})$ in $Form(\mathcal{L}^\circ)$, if $\exists \mathrm{x} \mathrm{A}(\mathrm{x}) \in \Sigma$ then there is some d such that $\mathrm{A}(\mathrm{d}) \in \Sigma$.

Lemma 5.4.2.

Suppose $\Sigma \subseteq Form(\mathcal{L})$ and Σ is consistent. Σ can be extended to some maximal consistent $\Sigma^* \subseteq Form(\mathcal{L}^\circ)$ such that Σ^* has the E-property.

Proof. Since $Form(\mathcal{L}^\circ)$ is countable, the subset of existential formulas of $Form(\mathcal{L})$ is countable. Now suppose

$$(1) \qquad \exists \mathrm{x} \mathrm{A}_1(\mathrm{x}),\ \exists \mathrm{x} \mathrm{A}_2(\mathrm{x}),\ \exists \mathrm{x} \mathrm{A}_3(\mathrm{x}), \ldots$$

is an arbitrary enumeration of them. Of course the x's in (1) may or may not be different symbols.

Construct an infinite sequence of $\Sigma_n \subseteq Form(\mathcal{L}^\circ)$ $(n \geq 0)$ as follows.

Let $\Sigma_0 = \Sigma$.

Take the first existential formula $\exists \mathrm{x} \mathrm{A}_1(\mathrm{x})$ from (1). Since $\exists \mathrm{x} \mathrm{A}_1(\mathrm{x})$ is finite in length, we can find some d which does not occur in it. Of course d does not occur in Σ_0, because $\Sigma_0 = \Sigma \subseteq Form(\mathcal{L})$. Form A(d) from A(x) and let

$$\Sigma_1 = \Sigma_0 \cup \{\exists \mathrm{x} \mathrm{A}_1(\mathrm{x}) \to \mathrm{A}_1(\mathrm{d})\}.$$

Suppose Σ_0, ..., Σ_n have been constructed. Take $\exists x A_{n+1}(x)$ from (1). Then we can find some d which does not occur in $\exists x A_{n+1}(x)$ nor in Σ_n. (Since we have an unlimited supply of new symbols and as at each stage only a finite number of them has been used, we can always find a fresh d availble for this purpose.) Let

$$\Sigma_{n+1} = \Sigma_n \cup \{\exists x A_{n+1}(x) \to A_{n+1}(d)\}.$$

We will prove by induction that Σ_n ($n \geq 0$) is consistent. Σ_0 is consistent by supposition. Suppose Σ_n is consistent but Σ_{n+1} is not. We have the following:

$$\begin{aligned}
&\Sigma_n \vdash \neg(\exists x A_{n+1}(x) \to A_{n+1}(d)),\\
&\Sigma_n \vdash \exists x A_{n+1}(x) \wedge \neg A_{n+1}(d),\\
&\Sigma_n \vdash \forall y(\exists x A_{n+1}(x) \wedge \neg A_{n+1}(y)),\\
&\Sigma_n \vdash \exists x A_{n+1}(x) \wedge \forall y \neg A_{n+1}(y),\\
&\Sigma_n \vdash \exists x A_{n+1}(x) \wedge \neg\exists y A_{n+1}(y),\\
&\Sigma_n \vdash \exists x A_{n+1}(x) \wedge \neg\exists x A_{n+1}(x),
\end{aligned}$$

which contradict the induction hypothesis that Σ_n is consistent. Hence Σ_{n+1} is consistent.

Let $\Sigma^\circ = \bigcup_{n \in N} \Sigma_n$. It can be easily proved that Σ° is consistent. By Lemma 5.3.5, Σ° can be extended to some maximal consistent $\Sigma^* \subseteq Form(\mathcal{L}^\circ)$. Finally we will prove that Σ^* has the E-property.

Suppose $\exists x A(x) \in Form(\mathcal{L}^\circ)$. By the above construction of Σ_0, Σ_1, Σ_2, ..., there is some d and k such that $\exists x A(x) \to A(d) \in \Sigma_k$ and accordingly

$$\exists x A(x) \to A(d) \in \Sigma^*. \tag{2}$$

Suppose $\exists x A(x) \in \Sigma^*$. By (2), the maximal consistency of Σ^*, and Lemma 5.3.5 [4], we have $A(d) \in \Sigma^*$. Hence Σ^* has the E-property. □

We will use the maximal consistent set Σ^* in Lemma 5.4.2 to construct a valuation. We take the set

$$T = \{t' \mid t \in Term(\mathcal{L}^\circ)\}$$

to be the domain. Actually T is the same as $Term(\mathcal{L}^\circ)$ except that t in $Term(\mathcal{L}^\circ)$ is written as t′ in T. Then a valuation v over domain T is constructed satisfying the following:

1) For any individual symbol a and free variable symbol u in $\mathcal{L}$ and any new free variable symbol d in $\mathcal{L}^\circ$, $a^v = a' \in T$, $u^v = u' \in T$, and $d^v = d' \in T$.
2) For any n-ary relation symbol F and any $t'_1, \ldots, t'_n \in T$, $\langle t'_1, \ldots, t'_n \rangle \in F^v$ iff $F(t_1, \ldots, t_n) \in \Sigma^*$.
3) For any n-ary function symbol f and any $t'_1, \ldots, t'_n \in T$, $f^v(t'_1, \ldots, t'_n) = f(t_1, \ldots, t_n)' \in T$.

The conventions stated above will be used throughout this section.

Lemma 5.4.3.
For any $t \in Term(\mathcal{L}^\circ)$, $t^v = t' \in T$.

Proof. By induction on the structure of t. □

Lemma 5.4.4.
For any $A \in Form(\mathcal{L}^\circ)$, $A^v = 1$ iff $A \in \Sigma^*$.

Proof. By induction on the structure of A.

Basis. A is an atom $F(t_1, \ldots, t_n)$. In this case the lemma is proved by 2) and Lemma 5.4.3.

Induction step. We distinguish seven cases: $A = \neg B$, $B \wedge C$, $B \vee C$, $B \to C$, $B \leftrightarrow C$, $\forall x B(x)$, or $\exists x B(x)$. For the five cases concerning the connectives, the proof is exactly the same as that for Lemma 5.3.6. We shall prove the lemma for the case of $A = \exists x B(x)$ and leave that of $A = \forall x B(x)$ to the reader.

Since A is an existential formula, we have the following:

$$
\begin{aligned}
& \exists x B(x) \in \Sigma^* \\
\Longrightarrow\ & B(d) \in \Sigma^* \text{ for some d} \quad \text{(by Lem 5.4.2)} \\
\Longrightarrow\ & B(d)^v = 1 \quad \text{(by ind hyp)} \\
\Longrightarrow\ & \exists x B(x)^v = 1.
\end{aligned}
$$

For the converse, suppose $\exists x B(x)^v = 1$. Form B(u) from B(x), where u does not occur in B(x). We have

(1) There exists $t' \in T$ (that is, $t \in Term(\mathcal{L}^\circ)$), such that $B(u)^{v(u/t')} = 1$.

Form B(t) from B(x). Since $t^v = t'$ (by Lemma 5.4.3), we have

(2) $B(t)^v = B(u)^{v(u/t^v)} = B(u)^{v(u/t')}$,

and then

$\exists x B(x)^v = 1$

$\Longrightarrow$ There exists $t' \in T$, such that $B(u)^{v(u/t')} = 1$ (by (1))

$\Longrightarrow B(t)^v = 1$ (by(2))

$\Longrightarrow B(t) \in \Sigma^*$ (by ind hyp)

$\Longrightarrow \Sigma^* \vdash B(t)$

$\Longrightarrow \Sigma^* \vdash \exists x B(x)$

$\Longrightarrow \exists x B(x) \in \Sigma^*$ (by Lem 5.3.2).

Thus the induction step is proved. □

Theorem 5.4.5. (*Completeness*)
Suppose $\Sigma \subseteq Form(\mathcal{L})$. If Σ is consistent, then Σ is satisfiable.

Proof. By Lemmas 5.4.2 and 5.4.4. □

Theorem 5.4.6. (*Completeness*)
Suppose $\Sigma \subseteq Form(\mathcal{L})$ and $A \in Form(\mathcal{L})$. Then

[1] If $\Sigma \models A$, then $\Sigma \vdash A$.

[2] If $\emptyset \models A$, then $\emptyset \vdash A$.
(That is, every valid formula is formally provable.) □

According to Gödel [1930], Σ is countable. Henkin [1949] extended Σ to uncountable sets.

$\Sigma \models A$ is said to hold in domain D if for every valuation v over D, $\Sigma^v = 1$ implies $A^v = 1$.

Since T is countably infinite, Theorems 5.4.5 and 5.4.6 can be stated more precisely as follows.

Theorem 5.4.5. (*Completeness*)
If Σ is consistent, then Σ is satisfiable in a countably infinite domain. Hence if Σ is consistent, Σ is satisfiable. □

Theorem 5.4.6. (*Completeness*)

[1] If $\Sigma \models A$ in a countably infinite domain, then $\Sigma \vdash A$. Hence if $\Sigma \models A$, then $\Sigma \vdash A$.

[2] If A is valid in a countably infinite domain, then A is formally provable. Hence if A is valid, A is formally provable. □

Remarks

In the construction of the valuation v, if the relation symbol in 2) were the equality symbol, the requirement would be

$$(1) \qquad t_1' = t_2' \quad \text{iff} \quad t_1 \approx t_2 \in \Sigma^*.$$

Suppose t_1 and t_2 are different terms, that is, $t_1' \neq t_2'$. Since Σ^* is constructed before v, it may be true that $t_1 \approx t_2 \in \Sigma^*$. Thus (1) may be false. For instance, suppose $t_1 = u$ and $t_2 = v$, where u and v are different free variable symbols. Then $u' \neq v'$. Let $\Sigma = \{u \approx v\}$. Obviously Σ is satisfiable. By Soundness Theorem, Σ is consistent. Hence Σ can be extended to maximal consistent Σ^*, and then $u \approx v \in \Sigma^*$.

Hence 2) is not available for proving the completeness of first-order logic with equality.

5.5. COMPLETENESS OF FIRST-ORDER LOGIC WITH EQUALITY

As mentioned in the last section, we first extend a given consistent $\Sigma \subseteq Form(\mathcal{L})$ to some maximal consistent $\Sigma^* \subseteq Form(\mathcal{L}^\circ)$ such that Σ^* has the E-property. We will still let $T = \{t' \mid t \in Term(\mathcal{L}^\circ)\}$.

In this section, however, the equality symbol is contained in $\mathcal{L}$ and $\mathcal{L}^\circ$.

We define a binary relation $\sim$ on $Term(\mathcal{L}^\circ)$ by

$$1) \qquad t_1 \sim t_2 \quad \text{iff} \quad t_1 \approx t_2 \in \Sigma^*.$$

By 1) we can prove that for any $t_1, t_2, t_3 \in Term(\mathcal{L}^\circ)$,

$$2) \qquad t_1 \sim t_1,$$

$$3) \qquad t_1 \sim t_2 \Longrightarrow t_2 \sim t_1,$$

$$4) \qquad t_1 \sim t_2 \text{ and } t_2 \sim t_3 \Longrightarrow t_1 \sim t_3.$$

The proof is left to the reader.

By 2)–4), $\sim$ is an equivalence relation. For every $t \in Term(\mathcal{L}^\circ)$, the $\sim$-equivalence class of t is

$$\bar{t} = \{t_1 \in Term(\mathcal{L}^\circ) \mid t \sim t_1\}.$$

We have $t \sim t_1$ iff $\bar{t} = \bar{t}_1$. (See Section 1.1.)

Let

$$\overline{T} = \{\bar{\mathrm{t}} \mid \mathrm{t} \in Term(\mathcal{L}^\circ)\}.$$

Then we have

5) $$0 < |\overline{T}| \leq |T|.$$

Thus $\overline{T}$ is (non-empty) finite or countably infinite, because T is countably infinite.

We want to prove that if $\mathrm{t}_i \sim \mathrm{t}_i^\circ$ $(i = 1, 2, \ldots, n)$, F and f are any n-ary relation symbol and function symbol, then

6) $$\mathrm{F}(\mathrm{t}_1, \ldots, \mathrm{t}_n) \in \Sigma^* \quad \text{iff} \quad \mathrm{F}(\mathrm{t}_1^\circ, \ldots, \mathrm{t}_n^\circ) \in \Sigma^*.$$

7) $$\mathrm{t}_1 \approx \mathrm{t}_2 \in \Sigma^* \quad \text{iff} \quad \mathrm{t}_1^\circ \approx \mathrm{t}_2^\circ \in \Sigma^*.$$

8) $$\mathrm{f}(\mathrm{t}_1, \ldots, \mathrm{t}_n) \sim \mathrm{f}(\mathrm{t}_1^\circ, \ldots, \mathrm{t}_n^\circ).$$

We shall prove 6) and leave the proof of 7) and 8) to the reader. Suppose $\mathrm{F}(\mathrm{t}_1, \ldots, \mathrm{t}_n) \in \Sigma^*$. We have

9) $$\Sigma^* \vdash \mathrm{F}(\mathrm{t}_1, \ldots, \mathrm{t}_n).$$

By the supposition $\mathrm{t}_i \sim \mathrm{t}_i^\circ$, and by 1) and the maximal consistency of Σ^* we have

10) $$\Sigma^* \vdash \mathrm{t}_i \approx \mathrm{t}_i^\circ \ (i = 1, \ldots, n).$$

By 9) and 10) we have $\Sigma^* \vdash \mathrm{F}(\mathrm{t}_1^\circ, \ldots, \mathrm{t}_n^\circ)$ and accordingly $\mathrm{F}(\mathrm{t}_1^\circ, \ldots, \mathrm{t}_n^\circ) \in \Sigma^*$. The converse will be proved similarly.

Now we use Σ^* to construct a valuation v over domain $\overline{T}$ satsifying the following:

11) $\mathrm{a}^v = \bar{\mathrm{a}} \in \overline{T}$; $\mathrm{u}^v = \bar{\mathrm{u}} \in \overline{T}$; $\mathrm{d}^v = \bar{\mathrm{d}} \in \overline{T}$.
12) For any $\bar{\mathrm{t}}_1, \ldots, \bar{\mathrm{t}}_n \in \overline{T}$, $\langle \bar{\mathrm{t}}_1, \ldots, \bar{\mathrm{t}}_n \rangle \in \mathrm{F}^v$ iff $\mathrm{F}(\mathrm{t}_1, \ldots, \mathrm{t}_n) \in \Sigma^*$, where t_i may be (by 6)) any member of $\bar{\mathrm{t}}_i$ $(i = 1, \ldots, n)$.
    For any $\bar{\mathrm{t}}_1, \bar{\mathrm{t}}_2 \in \overline{T}$, $\langle \bar{\mathrm{t}}_1, \bar{\mathrm{t}}_2 \rangle \in \approx^v$ (that is, $\bar{\mathrm{t}}_1 = \bar{\mathrm{t}}_2$) iff $\mathrm{t}_1 \approx \mathrm{t}_2 \in \Sigma^*$, where t_1 and t_2 may be (by 7)) any member of $\bar{\mathrm{t}}_1$ and $\bar{\mathrm{t}}_2$ respectively.
13) For any $\bar{\mathrm{t}}_1, \ldots, \bar{\mathrm{t}}_n \in \overline{T}$, $\mathrm{f}^v(\bar{\mathrm{t}}_1, \ldots, \bar{\mathrm{t}}_n) = \overline{\mathrm{f}(\mathrm{t}_1, \ldots, \mathrm{t}_n)} \in \overline{T}$, where t_i may be (by 8)) any member of $\bar{\mathrm{t}}_i$ $(i = 1, \ldots, n)$.

We will explain why t_i may be any member of $\bar{\mathrm{t}}_i$ $(i = 1, 2, \ldots, n)$ in 12) and 13).

Suppose $t_i^\circ \in \bar{t}_i$ $(i = 1, 2, \ldots, n)$. Then we have $t_i \sim t_i^\circ$, and accordingly 6)–8) and

$$\overline{f(t_1, \ldots, t_n)} = \overline{f(t_1^\circ, \ldots, t_n^\circ)}.$$

Hence t_i may be any member of $\bar{t}_i$ in 12) and 13).

Since $\sim$ is an equivalence relation on $Term(\mathcal{L}^\circ)$, hence, for the equality symbol $\approx$, given any $\bar{t}_1$, $\bar{t}_2 \in \overline{T}$, we have

$$\begin{aligned} &\langle \bar{t}_1, \bar{t}_2 \rangle \in \approx^v \quad (\text{that is}, \bar{t}_1 = \bar{t}_2) \\ \Longleftrightarrow\ & t_1 \sim t_2 \\ \Longleftrightarrow\ & t_1 \approx t_2 \in \Sigma^*. \end{aligned}$$

This means, when $t_1 \approx t_2 \in \Sigma^*$, althought t_1 and t_2 may be different members in *Term* $(\mathcal{L}^\circ)$ (that is, t_1' and t_2' are different individuals in T), yet since $t_1 \sim t_2$, $\bar{t}_1$ and $\bar{t}_2$ are the same individual in $\overline{T}$.

Lemma 5.5.1.
For any $t \in Term(\mathcal{L}^\circ)$, $t^v = \bar{t} \in \overline{T}$. □

Lemma 5.5.2.
For any $A \in Form(\mathcal{L}^\circ)$, $A^v = 1$ iff $A \in \Sigma^*$.

Proof. Analogous to the proof of Lemma 5.4.4. □

Since the domain $\overline{T}$ for valuation v is countably infinite or finite, we have the following completeness theorem for first-order logic with equality.

Theorem 5.5.3. (***Completeness***)
Suppose $\Sigma \subseteq Form(\mathcal{L})$, where $\mathcal{L}$ is the first-order language with equality. If Σ is consistent, then Σ is satisfiable in a countably infinite domain or in some finite domain. Hence if Σ is consistent, then Σ is satisfiable. □

Theorem 5.5.4. (***Completeness***)
Suppose $\Sigma \subseteq Form(\mathcal{L})$ and $A \in Form(\mathcal{L})$ with $\mathcal{L}$ being the first-order language with equality.

[1] If $\Sigma \models A$ in a countably infinite domain and in every finite domain, then $\Sigma \vdash A$. Hence if $\Sigma \models A$,then $\Sigma \vdash A$.

[2] If A is valid in a countably infinite domain and in every finite domain, then A is formally provable. Hence if A is valid, then A is formally provable. □

5.6. INDEPENDENCE

A rule of formal deduction is said to be *independent* iff it is not derivable from the remaining ones. An independent rule is indispensable, and a dependent one is redundant.

Although it seems natural to require each of the rules of formal deduction to be independent, we may preserve some dependent ones for certain reasons. Hence the requirement of independence is more of aesthetic significance than necessity.

Suppose a system of rules of formal deduction are given and (R) is one of them. Suppose each rule in the system other than (R) has either a certain property (if it generates schemes of formal deducibility directly) or preserves this property (if it generates schemes from given ones), and there is some scheme $\Sigma \vdash A$ of this system which does not have this property. Then $\Sigma \vdash A$ cannot be derived from the remaining rules and accordingly (R) is independent. This gives a general method of an independence proof.

We prove the independence of the rules of formal deduction of first-order logic as follows.

Case of (Ref). It is easy to see that each rule other than (Ref) has or perserves the following property:

1) Suppose $\Sigma \vdash A$ occurs in the rule. Replace Σ by $\emptyset$. Then the result $\emptyset \vdash A$ holds.

For instance, $(\approx +)$ obviously has this property. $(\forall +)$ preserves this property because $\emptyset \vdash A(u)$ implies $\emptyset \vdash \forall x A(x)$. But in the case of $F(u) \vdash F(u)$, $\emptyset \vdash F(u)$ does not hold becasuse $\emptyset \models F(u)$ does not hold. Hence $F(u) \vdash F(u)$ does not have the property 1), and accordingly (Ref) is independent.

Case of (+). Suppose (+) is deleted. In the remaining part, the only two rules (Ref) and $(\approx +)$, which generate schemes directly, have the property that there is at most one formula in the premises. The schemes generated by the other rules have the same or a smaller number of formulas in the premises than the given ones. Hence they preserve the above property. But in the scheme

$$A, B \vdash A$$

there are two formulas in the premises. Hence (+) is independent.

In the rules concerning the connectives, we shall prove the independence of $(\neg-)$ and $(\rightarrow+)$ while the rest will be left to the reader.

Case of $(\neg-)$. The value of $\neg$A has been defined by

$$(\neg A)^v = \begin{cases} 1 & \text{if } A^v = 0, \\ 0 & \text{otherwise.} \end{cases}$$

But here we stipulate that

$$(\neg A)^v = 1 \quad \text{for} \quad A^v = 1 \quad \text{or} \quad A^v = 0.$$

Then each rule other than $(\neg-)$ has or preserves the following property:

2) Suppose $\Sigma \vdash A$ occurs in the rule, and v is any valuation.
If $\Sigma^v = 1$, then $A^v = 1$.

This is obvious because this stipulation is concerned only with negation. But in the scheme

$$\neg\neg A \vdash A \tag{3}$$

we may set $A^v = 0$ and have $(\neg\neg A)^v = 1$ such that 3) does not have the property 2). Hence $(\neg-)$ is independent.

Case of $(\rightarrow+)$. We make the stipulation

$$(A \rightarrow B)^v = 0 \quad \text{if} \quad A^v = 1. \tag{4}$$

Then each rule not concerning implication will have or preserve the property 2). For the rule $(\rightarrow-)$:

$$\begin{aligned} \text{If} \quad & \Sigma \vdash A \rightarrow B, \\ & \Sigma \vdash A, \\ \text{then} \quad & \Sigma \vdash B. \end{aligned}$$

we suppose

$$\Sigma^v = 1 \Longrightarrow (A \rightarrow B)^v = 1$$

and

$$\Sigma^v = 1 \Longrightarrow A^v = 1\,.$$

Then $\Sigma^v \neq 1$. (If $\Sigma^v = 1$, we obtain $(A \rightarrow B)^v = 1$ and $A^v = 1$, contradicting 4).) Accordingly we have

$$\Sigma^v = 1 \Longrightarrow B^v = 1\,.$$

Hence $(\to -)$ also preserves the property 2). Thus each rule other than $(\to +)$ has or preserves the property 2). But in the scheme

5) $$A \vdash B \to A$$

we may let $A^v = B^v = 1$ and have $(B \to A)^v = 0$ such that 5) does not have the property 2). Hence $(\to +)$ is independent.

Case of $(\forall -)$. Suppose A' results from A by replacing each segment $\forall xB$ in A by $\forall x(B \to B)$. For instance, if $A = \forall xyF(x, y)$, then

$$A' = \forall x[\forall y(F(x, y) \to F(x, y)) \to \forall y(F(x, y) \to F(x, y))].$$

Suppose $\Sigma' = \{A' \mid A \in \Sigma\}$. Then each rule not concerning $\forall$ has or preserves the following property:

6) Suppose $\Sigma \vdash A$ occurs in the rule. Then $\Sigma' \vdash A'$ holds.

This is because the above replacement of A by A' is not involved in such a rule. For $(\forall +)$:

$$\begin{aligned} &\text{If} \quad \Sigma \vdash A(u), u \text{ not occurring in } \Sigma, \\ &\text{then} \quad \Sigma \vdash \forall xA(x). \end{aligned}$$

the resulting scheme $\Sigma \vdash \forall xA(x)$ in it after the replacement becomes

$$\Sigma' \vdash \forall x(A(x)' \to A(x)')$$

which obviously holds. Hence $(\forall +)$ preserves the property 6). Thus each rule other than $(\forall -)$ has or preserves the property 6). But the scheme

7) $$\forall xF(x) \vdash F(u)$$

after replacement becomes

$$\forall x(F(x) \to F(x)) \vdash F(u)$$

which does not hold because

$$\forall x(F(x) \to F(x)) \models F(u)$$

does not hold. Hence 7) does not have the property 6), which proves the independence of $(\forall -)$.

The independence of $(\forall+)$, $(\exists-)$, and $(\exists+)$ can be proved in a similar way as that for $(\forall-)$, with suitable modifications.

Case of $(\approx-)$. Suppose A' results from A by replacing each atom $t_1 \approx t_2$ in A by

$$t_1 \approx t_2 \rightarrow t_1 \approx t_2$$

and suppose $\Sigma' = \{A' \mid A \in \Sigma\}$. Then each rule which does not concern the equality symbol has or preserves the property 6). $(\approx+)$ after replacement becomes

$$\emptyset \vdash u \approx u \rightarrow u \approx u$$

which obviously holds. Then each rule other than $(\approx-)$ has or preserves the property 6). But the scheme

8) $$F(u), u \approx v \vdash F(v)$$

after replacement becomes

$$F(u), u \approx v \rightarrow u \approx v \vdash F(v)$$

which does not hold because

$$F(u), u \approx v \rightarrow u \approx v \models F(v)$$

does not hold. Hence 8) does not have the property 6) and accordingly $(\approx-)$ is independent.

The independence of $(\approx+)$ can be proved in a similar way with suitable modifications.

Now we turn to consider the independence of the axioms in the system of formal deducibility of another type described in Chapter 4. They will be proved, essentially in an analogous way.

For simplicity, we will consider the subsystem of propositional logic based upon negation and implication, which form an adequate set of connectives (see Section 2.8). The three axioms are:

(Ax1) $A \rightarrow (B \rightarrow A)$

(Ax2) $(A \rightarrow (B \rightarrow C)) \rightarrow ((A \rightarrow B) \rightarrow (A \rightarrow C))$

(Ax3) $(\neg A \rightarrow B) \rightarrow ((\neg A \rightarrow \neg B) \rightarrow A)$

and the one rule of inference is

(R1) From $A \rightarrow B$ and A infer B.

The truth table of implication:

A	B	A → B
1	1	1
1	0	0
0	1	1
0	0	1

can be written in a simpler form and combined with the truth table of negation as follows:

→	1	0	¬
1	1	0	0
0	1	1	1

For the proof of independence, more values are adopted instead of the original truth and falsehood. Here four values 0, 1, 2, and 3 are adopted, which are not intended to denote truth or falsehood. New "truth tables" for negation and implication are then stipulated as follows:

→	0	1	2	¬
0	0	2	2	2
1	2	2	0	0
2	0	0	0	0

It can be verified that, according to these tables, (Ax2) and (Ax3) have the following property:

9) The whole formula always has the value 0 for any values of 0, 1, 2 assigned to A, B, C in it.

and (R1) preserves this property. (The verification is left to the reader.) But if 0 and 1 are assigned to A and B respectively, (Ax1) will have the value 2. Hence (Ax1) does not have the property 9), which proves its independence.

For the independence of (Ax2), the following truth tables:

$\to$	0	1	2	3	$\neg$
0	0	1	1	3	3
1	0	0	1	0	0
2	0	0	0	3	0
3	0	0	0	0	0

are constructed, according to which (Ax1) and (Ax3) have the property 9), and (R1) preserves it. (The verification is left to the reader.) But (Ax2) will have the value 1 when 1, 1, 2 are assigned, respectively, to A, B, C. Hence (Ax2) is independent.

For the independence of (Ax3), we construct the following truth tables:

$\to$	0	1	$\neg$
0	0	1	0
1	0	0	0

according to which (Ax1) and (Ax2) have the property 9), and (R1) preserves it. (The verification is left to the reader.) But (Ax3) will have the value 1 when 1 and 0 are assigned, respectively, to A and B. Hence (Ax3) is independent.

Finally the rule (R1) is independent because no formulas of forms other than those of the axioms can be derived without it.

Exercises 5.6.

5.6.1. Complete the proof of independence of the rules of formal deduction of first-order logic.

5.6.2. Prove that ($\neg-$) in the rules of formal deduction cannot be replaced by ($\neg+$), as mentioned in Section 2.6 of Chapter 2.

5.6.3. Prove the independence of the axioms of the following system of propositional logic:

Axioms:

(1) $A \to (B \to A)$

(2) $(A \to (B \to C)) \to ((A \to B) \to (A \to C))$

(3) $(\neg A \to \neg B) \to (B \to A)$

The rule of inference is (R1).

6

COMPACTNESS, LÖWENHEIM-SKOLEM AND HERBRAND THEOREMS

Many important results can be obtained by applying the Soundness and Completeness Theorems, among which are Compactness, Löwenheim–Skolem's, and Herbrand's Theorems.

6.1. COMPACTNESS THEOREM

Theorem 6.1.1. (*Compactness*)
$\Sigma \subseteq Form(\mathcal{L})$ is satisfiable iff every finite subset of Σ is satisfiable.

Proof. Suppose every finite subset of Σ is satisfiable. By the Soundness Theorem, every finite subset of Σ is consistent. If Σ is inconsistent, then some finite subset of Σ is inconsistent, yielding a contradiction. Hence Σ is consistent. By the Completeness Theorem, Σ is satisfiable.
The converse is obvious. □

Corollary 6.1.2.
If $\Sigma \subseteq Form(\mathcal{L})$ is satisfiable in any finite domain, then Σ is satisfiable in an infinite domain.

Proof. Suppose Σ is satisfiable in any finite domain. Let

$$\mathcal{D} = \{\mathrm{d}_0, \mathrm{d}_1, \mathrm{d}_2, \dots\}$$

be some countable set of new free variable symbols, that is, $\mathcal{D}$ and $\mathcal{L}$ are disjoint. Consider the set Σ' defined by

$$\Sigma' = \Sigma \cup \{\neg(\mathrm{d}_m \approx \mathrm{d}_n) \mid m < n\}.$$

Any finite subset Σ° of Σ' will involve at most $\mathrm{d}_0, \ldots, \mathrm{d}_k$ of $\mathcal{D}$, say, for some k. The formulas $\neg(\mathrm{d}_m \approx \mathrm{d}_n)$ in Σ° are satisfiable in any domain with at least $k+1$ elements. By supposition, Σ is also satisfiable in this domain. Since $\mathrm{d}_0, \ldots, \mathrm{d}_k$ do not occur in Σ, the two parts of Σ° can be satisfied simultaneously by the same valuation over this domain.

By the Compactness Theorem, Σ' is satisfiable. Since any finite domain is not available for this case, Σ' must be satisfiable in an infinite domain, and so is Σ. □

Exercises 6.1.

6.1.1. Suppose $\Sigma \models \mathrm{A}$. Then $\Delta \models \mathrm{A}$ for some finite $\Delta \subseteq \Sigma$. (Not use Theorem 2.6.2.)

6.1.2. Suppose $\Sigma \subseteq Form(\mathcal{L})$ does not contain equality symbol and D is an infinite domain. If for each valuation v over D there is $\mathrm{A} \in \Sigma$ such that $\mathrm{A}^v = 1$, then there are $\mathrm{B}_1, \ldots, \mathrm{B}_k \in \Sigma$ such that $\mathrm{B}_1 \vee \cdots \vee \mathrm{B}_k$ is valid.

6.1.3. In Exercise 6.1.2, suppose D is finite and delete the supposition that Σ does not contain equality symbol, then $\mathrm{B}_1 \vee \cdots \vee \mathrm{B}_k$ is valid in D.

6.2. LÖWENHEIM–SKOLEM'S THEOREM

Theorem 6.2.1. (***Löwenheim–Skolem***)

Suppose

$\Sigma \subseteq Form(\mathcal{L})$.

[1] Σ not containing equality is satisfiable iff Σ is satisfiable in a countably infinite domain.

[2] Σ containing equality is satisfiable iff Σ is satisfiable in a countably infinite domain or in some finite domain.

Proof. By the Soundness and Completeness Theorems. □

Theorem 6.2.1 was first proved by Löwenheim [1915] for finite Σ, but the proof had several gaps. Skolem [1920] established the complete proof of the theorem and extended Σ to countable sets.

In this book only countable sets are considered. Hence the formal languages and the domain of valuations are at most countably infinite. Löwenheim–Skolem's Theorem will have stronger forms (downward and upward Löwenheim–Skolem Theorems) if sets of any transfinite cardinality are considered.

Löwenheim–Skolem's Theorem can be formulated in terms of validity.

Theorem 6.2.2. ***(Löewenheim–Skolem)***
Suppose $A \in Form(\mathcal{L})$.

[1] A not containing equality is valid iff A is valid in a countably infinite domain.

[2] A containing equality is valid iff A is valid in a countably infinite domain and in every finite domain. □

6.3. HERBRAND'S THEOREM

Herbrand's Theorem is the basis of one of the approaches of automatic theorem proving in artificial intelligence.

In order to formulate Herbrand's Theorem, we have to begin with some preliminary definitions and theorems.

First of all we shall transform a prenex normal form to an ∃-free prenex normal form by deleting the existential quantifiers. Suppose ∃y is the left-most existential quantifier in a prenex normal form A. When no universal quantifier occurs on the left of ∃y, we use any free variable symbol u not occuring in A or in this procedure, and substitute u for (all occurrences of) y in the matrix of A. If $\forall x_1, \ldots, \forall x_n$ occur in this order on the left of ∃y, we use any n-ary function symbol f not occurring in A or in this procedure, and substitute $f(x_1, \ldots, x_n)$ for y in the matrix of A. Then ∃y is deleted. The formula, which results after deleting all the existential quantifiers in a prenex normal form, is called an *∃-free prenex normal form* of the original formula.

For instance, let

$$A = \exists y_1 y_2 \forall x_1 \exists y_3 \forall x_2 x_3 \exists y_4 y_5 \forall x_4 \\ B(y_1, y_2, x_1, y_3, x_2, x_3, y_4, y_5, x_4).$$

Use free variable symbols u and v, unary function symbol f, and ternary function symbols g and h, which do not occur in A. Then

$$\forall x_1 x_2 x_3 x_4 B(u, v, x_1, f(x_1), x_2, x_3, \\ g(x_1, x_2, x_3), h(x_1, x_2, x_3), x_4)$$

is an $\exists$-free prenex normal form of A. It may be written simply in the form $\forall x_1 x_2 x_3 x_4 B'(x_1, x_2, x_3, x_4)$.

Theorem 6.3.1.
A prenex normal form A is satisfiable in a domain D iff its $\exists$-free prenex normal form is satisfiable in D; hence A is satisfiable iff its $\exists$-free prenex normal form is satisfiable.

Proof. To prove this theorem, we may suppose without loss of generality

$$A = \exists x \forall y \exists z B(x, y, z).$$

Then the $\exists$-free prenex normal form of A is

$$\forall y B(u, y, f(y)) \tag{1}$$

where u and f do not occur in A.

By Theorem 5.1.2, A is satisfiable in D iff

$$\forall y \exists z B(u, y, z) \tag{2}$$

is satisfiable in D. We want to prove that (2) is satisfiable in D iff (1) is satisfiable in D.

Suppose (1) is satisfiable by a valuation v over D. Then $\forall y\, B(u, y, f(y))^v = 1$, that is, for every $\alpha \in D$,

$$B(u, v, f(v))^{v(v/\alpha)} = 1 \tag{3}$$

where v does not occur in $B(u, y, f(y))$. We have $f(v)^{v(v/\alpha)} = f^v(\alpha) \in D$.

Since

$$f(v)^{v(v/\alpha)} = f^{v}(\alpha) = w^{v(v/\alpha)(w/f^{v}(\alpha))}, \tag{4}$$

we obtain by (3) and (4)

$$B(u, v, w)^{v(v/\alpha)(w/f^{v}(\alpha))} = B(u, v, f(v))^{v(v/\alpha)} = 1$$

where w does not occur in $B(u, v, f(v))$. Accordingly

$$\exists z B(u, v, z)^{v(v/\alpha)} = 1.$$

Then $\forall y \exists z B(u, y, z)^{v} = 1$ and (2) is satisfiable in D.

For the converse, suppose $\forall y \exists z B(u, y, z)^{v} = 1$, that is, for every $\alpha \in D$, there is some $\beta \in D$ such that

$$B(u, v, w)^{v(v/\alpha)(w/\beta)} = 1, \tag{5}$$

where v and w do not occur in $B(u, y, z)$. Let v' be any valuation over D such that v' agrees with v except that $f^{v'}(\alpha) = \beta$. By (5) we have

$$B(u, v, w)^{v'(v/\alpha)(w/f^{v'}(\alpha))} = 1. \tag{6}$$

Since

$$f(v)^{v'(v/\alpha)} = f^{v'}(\alpha) = w^{v'(v/\alpha)(w/f^{v'}(\alpha))}, \tag{7}$$

we obtain by (6) and (7)

$$B(u, v, f(v))^{v'(v/\alpha)} = B(u, v, w)^{v'(v/\alpha)(w/f^{v'}(\alpha))} = 1,$$

and accordingly $\forall y B(u, y, f(y))^{v'} = 1$. Hence (1) is satisfiable in D. □

By definition, a formula A is unsatisfiable iff it is false under all valuations over all domains. Since it is inconvenient and impossible to consider all valuations over all domains, it would be of great help if we could fix on some special domain such that A is unsatisfiable iff it is false under all valuations over this domain. Indeed, for any formula A, there does exist such a domain, which is the Herbrand universe of A.

Definition 6.3.2. (***Herbrand universe***)
Suppose A is an $\exists$-free prenex normal form. The set

{t′ | t is a term generated from the individual symbols,
free variable symbols, and function symbols occurring
in A. (If no individual symbol or free variable symbol
occurs in A, an arbitrary free variable symbol is to
be used.)}

is called the *Herbrand universe* of A and is denoted by H_A or simply by H.

Example.

If $A = \forall x(F(u) \wedge F(b) \wedge F(f(x)))$, then

$$H = \{u, b, f(u), f(b), f(f(u)), f(f(b)), \dots\}.$$

If $A = \forall x(F(f(u)) \vee G(b, g(x)))$, then

$$H = \{u, b, f(u), f(b), g(u), g(b), f(f(u)), f(f(b)), f(g(u)), f(g(b)), g(f(u)), g(f(b)), g(g(u)), g(g(b)), \dots\}.$$

If $A = \forall xy(F(x) \vee G(x, y))$, then $H = \{u\}$, where u is an arbitrary free variable symbol.

Definition 6.3.3. (***Herbrand valuation***)

Given an $\exists$-free prenex normal form A. A valuation v over the Herbrand universe H of A is called a *Herbrand valuation* if it satisfies [1] and [2]:

[1] $a^v = a' \in H$,
$u^v = u' \in H$.

[2] For any $t'_1, \dots, t'_n \in H, f^v(t'_1, \dots, t'_n) = f(t_1, \dots, t_n)' \in H$.

where a, u, f are, respectively, any individual symbol, free variable symbol, and n-ary function symbol occurring in A (u may be used arbitrarily in H).

Obviously $t^v = t' \in H$ for any Herbrand valuation v and any term t described in Definition 6.3.2.

Theorem 6.3.4.

An $\exists$-free prenex normal form A is unsatisfiable iff A is false under all Herbrand valuations.

Proof. It is obvious that the unsatisfiability of A implies its falsehood under all Herbrand valuations.

For the converse, suppose A is false under all Herbrand valuations, and suppose A is satisfiable, that is, there is some valuation v' over domain D such that $A^{v'} = 1$. We construct a Herbrand valuation v such that, besides the conditions in Definition 6.3.3, v satisfies

(1) For any n-ary relation symbol F occurring in A and any $t'_1, \dots, t'_n \in H, \langle t'_1, \dots, t'_n \rangle \in F^v$ iff $\langle t_1^{v'}, \dots, t_n^{v'} \rangle \in F^{v'}$, that is, $F(t_1, \dots, t_n)^v = F(t_1, \dots, t_n)^{v'}$.
Similarly for the equality symbol : $(t_1 \approx t_2)^v = (t_1 \approx t_2)^{v'}$.

We want to prove $A^v = 1$.

For any atom C, it is obvious by (1) that $C^v = C^{v'}$. Accordingly,

(2) For any quantifier-free A, $A^v = A^{v'} = 1$.

If A contains quantifiers, we may without loss of generality suppose $A = \forall x B(x)$, where B(x) is quantifier-free. Take any $t' \in H$; then $t^{v'} \in D$. Since $A^{v'} = \forall x B(x)^{v'} = 1$, we have

(3) $$B(u)^{v'(u/t^{v'})} = 1$$

where u does not occur in B(x) or t. Since B(u) is quantifier-free, we have by (2) and (3):

$$B(u)^{v(u/t')} = B(u)^{v(u/t^v)} = B(t)^v = B(t)^{v'} = B(u)^{v'(u/t^{v'})} = 1,$$

and accordingly $A^v = \forall x B(x)^v = 1$.

Thus $A^v = 1$ is proved by induction, contradicting the assumption that A is false under all Herbrand valuations. Hence A is unsatisfiable. □

Suppose an ∃-free prenex normal form $\forall x_1, \ldots, x_n B(x_1, \ldots, x_n)$ be given. By an *instance* of the matrix $B(x_1, \ldots, x_n)$ we mean a formula $B(t_1, \ldots, t_n)$ obtained from the matrix by substitution, where $t_1, \ldots, t_n$ are any terms such that $t'_1, \ldots, t'_n$ are elements of the Herbrand universe of the given formula.

Theorem 6.3.5. (***Herbrand***)

An ∃-free prenex normal form $\forall x_1 \ldots x_n B(x_1, \ldots, x_n)$ is unsatisfiable iff there are finitely many instances of the matrix which are unsatisfiable.

Proof. We may, without loss of generality, suppose the given formula is $\forall x B(x)$, where B(x) is quantifier-free.

Suppose $B(t_1)$, …, $B(t_n)$ are any finitely many instances of B(x). We have

$$\forall x B(x) \vdash B(t_1) \wedge \ldots \wedge B(t_n).$$

Then the satisfiability of $\forall x B(x)$ implies that of $B(t), \ldots, B(t_n)$. Hence if there exist finitely many instances of B(x) which are unsatisfiable, $\forall x B(x)$ is unsatisfiable.

For the converse, suppose $\forall x B(x)$ is unsatisfiable, and any finitely many instances of B(x) are satisfiable. By the Compactness Theorem, the set

$$\{B(t) | t' \in H\}$$

of all instances of B(x) is satisfiable, where H is the Herbrand universe of $\forall$xB(x). That is, there is some valuation v' such that, for any $t' \in H$,

(1) $$B(t)^{v'} = 1.$$

Since B(t) is quantifier-free, we can construct a Herbrand valuation v (see the proof of Theorem 6.3.4) such that $B(t)^v = B(t)^{v'} = 1$ by (1).

Take any $t' \in H$ and u which does not occur in B(x) or t. Since $t^v = t'$, we have

$$B(u)^{v(u/t')} = B(u)^{v(u/t^v)} = B(t)^v = 1,$$

and accordingly $\forall xB(x)^v = 1$, contradicting the assumption that $\forall$xB(x) is unsatisfiable. Hence there exist finitely many instances of B(x) which are unsatisfiable. □

Remarks

Herbrand Theorem reduces the unsatisfiability of $\forall$xB(x) to that of finitely many formulas, hence it becomes the basis of one approach of automatic theorem proving in artificial intelligence.

The proposition to be proved can be expressed as

$$A_1, \ldots, A_n \vdash A$$

or equivalently

$$\emptyset \vdash A_1 \wedge \ldots \wedge A_n \to A.$$

Then the question is to prove the unsatisfiability of

1) $$\neg(A_1 \wedge \ldots \wedge A_n \to A).$$

1) can be transformed into a prenex normal form, and further transformed into an $\exists$-free prenex normal form. Then the question is to prove the unsatisfiability of an $\exists$-free prenex normal form. By Herbrand Theorem, the question becomes to find a finite number of instances of its matrix which are unsatisfiable.

The generation of instances is to substitute the terms in the Herbrand universe of the $\exists$-free prenex normal form for the bound variable symbols in the matrix. Such terms can be classified into different degrees of complexity, according to the number of occurrences of function symbols in them. Substitution of terms of higher degree of complexity generates more instances, hence it is more possible that the instances generated are unsatisfiable. Then the proposition is proved.

For each stage before the unsatisfiable finitely many instances are obtained, there are two possibilities. The one is that they have not yet been obtained in that stage, and more instances need to be generated. The other is that such unsatisfiable finitely many instances do not exist at all, that is, the proposition does not hold.

Hence, the above approach is not a decision procedure, which requires to decide whether the proposition holds or not, and to give a proof in case it holds. The above approach is usually called a semi-decision procedure.

7
CONSTRUCTIVE LOGIC

Non-classical logics are to be introduced in the present and following chapters. Roughly non-classical logics can be divided into two groups, those that rival classical logics and those which extend it. We choose to introduce constructive logic of the first group and modal logic of the second in this book.

Constructive logic is the logic for constructive reasoning. Statements in constructive reasoning are interpreted in a constructive way. This chapter is a brief introduction to constructive logic, and the resulting system will be compared with the classical system developed in Chapters 2–5.

7.1. CONSTRUCTIVITY OF PROOFS

Existential statements in mathematics can be interpreted in different ways. For instance, the following statement

1) For any natural number n, there exists a prime greater than n.

can be interpreted in the usual sense of "existence", or as

2) For any natural number n, a prime greater than n can be found.

The claim by 2) is that a certain construction can be made. In proving this claim we have to find (construct) some particular prime number which is greater than n, while in proving 1) in the usual interpretation we need not make such a construction.

The prime greater than n can be constructed as follows. Find $n! + 1$ from n, and find the least prime p which divides $n! + 1$. p does not divide $n!$, hence p is greater than n. Then p is the required prime.

The interpretation of 1) as 2) and its proof are constructive, while the other kind of interpretation and proof mentioned above are non-constructive. Obviously a constructive interpretation conveys more information and a constructive proof requires more effort than a classical and non-constructive one. Hence from the constructive point of view, certain arguments in classical proofs are not acceptable. A typical example is the proof of the following statement:

3) There are irrational numbers a and b such that a^b is rational.

A classical proof can be given as follows. $(\sqrt{2})^{\sqrt{2}}$ is either rational or irrational. In the first case, we may take $a = b = \sqrt{2}$; in the second case, we may take $a = (\sqrt{2})^{\sqrt{2}}$ and $b = \sqrt{2}$, since then $a^b = 2$. This proof is classical because it does not determine which of the two cases holds and accordingly it does not actually construct the required a and b.

The clause "$(\sqrt{2})^{\sqrt{2}}$ is either rational or irrational" is an example of the law of excluded middle

4) $\mathcal{A}$ or not-$\mathcal{A}$.

From the classical point of view, 4) is valid since one of $\mathcal{A}$ and not-$\mathcal{A}$ holds. But a constructive proof of " $\mathcal{A}$ or $\mathcal{B}$" consists of specifying a proof of $\mathcal{A}$ or a proof of $\mathcal{B}$, therefore 4) is not necessarily valid from the constructive viewpoint.

Suppose $\mathcal{A}$ is "There is some element in D having the property R" and not-$\mathcal{A}$ is "Every element in D does not have the property R". Let R be a property such that, for every element in D, we can determine whether it has the property R or not. Then, if D is finite, we can examine every element of D and either verify $\mathcal{A}$ or verify not-$\mathcal{A}$. But if D is infinite, such verification is no longer possible. Hence, from the constructive viewpoint, the law of excluded middle is not acceptable for infinite sets.

One more example to show the distinction between these two kinds of reasoning. In order to prove

5) There is some element in D having the property R.

we may assume every element in D does not have the property R and deduce a contradiction. By *reductio ad absurdum*, we have

6) Not that every element in D does not have the property R.

Then, the classical reasoning, but not the constructive, allows 5) to be deduced from 6).

In the foregoing we have explained the basic distinction between constructive and classical reasoning. We will, however, not attempt to discuss the philosophical background of these viewpoints.

The logic for constructive reasoning is the constructive logic, which is distinct from the classical logic studied in Chapters 2 to 5.

7.2. SEMANTICS

The languages for constructive propositional and first-order logic are the same as $\mathcal{L}^p$ and $\mathcal{L}$ respectively. Hence the definitions of term, atom, formula, and sentence remain unchanged. But both the semantics and formal deduction for constructive logic are different from those for the classical.

In classical logics we have an intended interpretation for the formal language and the truth values of formulas. According to the intended interpretation, we establish the formal deduction rules which coincide with informal reasoning. But the case with constructive logic is quite different. The formal deduction rules in constructive logic is obtained by weakening the rule $(\neg-)$ in classical logic, which is regarded not acceptable. The semantics of constructive logic is established later.

There are various kinds of semantics for constructive logic. What we shall introduce here is due to Kripke, which is fairly simple.

We first treat semantics for constructive propositional logic. We will give some intuitive explanations before the definition.

In Chapters 2, 3, and 5, truth valuation (denote by t) for $\mathcal{L}^p$ was distinguished from valuation (denoted by v) for $\mathcal{L}$. From now on, for the convenience of description and notation, truth valuation will also be called valuation and denoted by v. In the meantime, tautology will be called valid formula and tautological consequence will be called logical consequence. Of course the distinction between the corresponding terms should be kept in mind.

In classical logic each valuation determines the value of a formula. But now what determines the value of a formula is not a single valuation, but a set of valuations, some of which are regarded as situated in a succession of time. Suppose A is a formula and v is a valuation. In classical logic $A^v = 1$ means that A is assigned truth by v and $A^v = 0$ means that A is assigned falsehood by it. But the situation is not the same in the constructive sense. Now $A^v = 1$ means that A has been assigned truth by v, and hence it is supposed to imply that A will be assigned truth by every valuation occurring later in the succession of time. $A^v = 0$ means that A has not yet been assigned truth by v (but not that A has been assigned falsehood by v), and A may be assigned truth by some valuation occurring later. Hence we note that 1 and 0 do not denote truth and falsehood in the constructive sense. We will illustrate the above ideas by an example.

Suppose $v_1, \ldots, v_5$ are five valuations and p, q, r are atoms. In the following diagram:

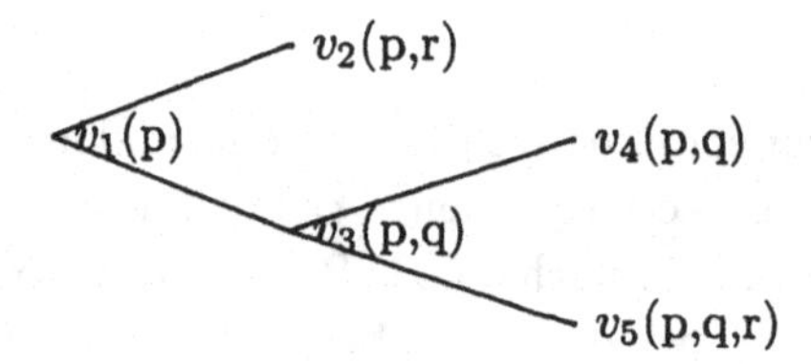

we have written some atom at a valuation v if v assigns the value 1 to this atom. We omit it at v if v assigns the value 0 to it. For instance, $p^{v_1} = 1$, $q^{v_1} = r^{v_1} = 0$. The diagram shows that v_1 precedes v_2 and v_3, and v_3 precedes v_4 and v_5. From v_1 we may proceed to v_2 or v_3. They are not identical because $q^{v_2} = 0$, $r^{v_2} = 1$, $q^{v_3} = 1$, $r^{v_3} = 0$. From v_3 we may proceed to v_4 or v_5. v_4 seems to be like v_3, but they are different. In fact, when we are at v_3, it is possible for us some time or other to proceed to v_5 and obtain $r^{v_5} = 1$. However, if we are at v_4, we will not be able to have the value 1 assigned to r.

Now we come to the definitions of a constructive valuation for $\mathcal{L}^p$, and the value of formulas under such valuations.

Definition 7.2.1. (***Constructive valuation for*** $\mathcal{L}^p$)

Suppose K is a set, R is a reflexive and transitive binary relation on K. Each $v \in K$ is called a *constructive valuation* for $\mathcal{L}^p$ which is a function with the set of all proposition symbols as domain and $\{1, 0\}$ as range, and which satisfies the condition that, for every proposition symbol p and every $v' \in K$, if $p^v = 1$ and vRv', then $p^{v'} = 1$.

Definition 7.2.2. (*Value of formulas*)

Suppose K and R are given as in Definition 7.2.1. The *value* of formulas under $v \in K$ is defined by recursion:

[1] $\mathrm{p}^v \in \{1, 0\}$.

[2] $(\mathrm{A} \wedge \mathrm{B})^v = \begin{cases} 1 & \text{if } \mathrm{A}^v = \mathrm{B}^v = 1, \\ 0 & \text{otherwise.} \end{cases}$

[3] $(\mathrm{A} \vee \mathrm{B})^v = \begin{cases} 1 & \text{if } \mathrm{A}^v = 1 \text{ or } \mathrm{B}^v = 1, \\ 0 & \text{otherwise.} \end{cases}$

[4] $(\mathrm{A} \to \mathrm{B})^v = \begin{cases} 1 & \text{if for every } v' \in K \text{ such that } vRv',\ \mathrm{A}^{v'} = 1 \\ & \text{implies } \mathrm{B}^{v'} = 1, \\ 0 & \text{otherwise.} \end{cases}$

[5] $(\mathrm{A} \leftrightarrow \mathrm{B})^v = \begin{cases} 1 & \text{if for every } v' \in K \text{ such that } vRv',\ \mathrm{A}^{v'} = \mathrm{B}^{v'}, \\ 0 & \text{otherwise.} \end{cases}$

[6] $(\neg \mathrm{A})^v = \begin{cases} 1 & \text{if for every } v' \in K \text{ such that } vRv',\ \mathrm{A}^{v'} = 0, \\ 0 & \text{otherwise.} \end{cases}$

Remarks

Constructively, Definition 7.2.2 does not work, since it clearly appeals to the law of excluded middle in clauses [4]–[6] (for instance, in [6], either for every v', $\mathrm{A}^{v'} = 0$ or not).

The forthcoming Definition 7.2.5 and proofs of Soundness and Completeness Theorems are also non-constructive.

Now we turn to treat semantics for constructive first-order logic.

Definition 7.2.3. (*Constructive valuation for $\mathcal{L}$*)

Suppose K and R are given as in Definition 7.2.1. Each $v \in K$ is called a *constructive valuation* for $\mathcal{L}$, which consists of a domain $D(v)$ which is peculiar to v and a function (denoted by v) with the set of all non-logical symbols, the equality symbol, and free variable symbols as domain such that for any individual symbol a, free variable symbol u, n-ary relation symbol F and function symbol f, the following, [1]–[4], are satisfied:

[1] If v, $v' \in K$ and vRv', then $D(v) \subseteq D(v')$.

[2] a^v, $\mathrm{u}^v \in D(v)$. If $v, v' \in K$ and vRv', then $\mathrm{a}^v = \mathrm{a}^{v'}, \mathrm{u}^v = \mathrm{u}^{v'}$.

[3] $F^v \subseteq D(v)^n$. If v, $v' \in K$ and vRv', then $F^v \subseteq F^{v'}$.
$\approx^v \subseteq D(v)^2$. If v, $v' \in K$ and vRv', then $\approx^v \subseteq \approx^{v'}$ obviously.

[4] $f^v: D(v)^n \to D(v)$. If $v, v' \in K$ and vRv', then $f^v = f^{v'}|D(v)$.

Definition 7.2.4. (***Value of terms***)

Suppose K and R are given as in Definition 7.2.1. The *value* of terms under $v \in K$ is defined by recursion:

[1] a^v, $u^v \in D$.

[2] $f(t_1, \ldots, t_n)^v = f^v(t_1^v, \ldots, t_n^v)$.

Theorem 7.2.5.

Suppose K and R are given, and $v \in K$. For any $t \in Term(\mathcal{L})$, $t^v \in D(v)$. If $v, v' \in K$ and vRv', then $t^v = t^{v'}$. □

Definition 7.2.6. (***Value of formulas***)

Suppose K and R are given as in Definition 7.2.1. The *value* of formulas under $v \in K$ is defined by recursion:

[1] $F(t_1, \ldots, t_n)^v = \begin{cases} 1 & \text{if } \langle t_1^v, \ldots, t_n^v \rangle \in F^v, \\ 0 & \text{otherwise.} \end{cases}$

$$(t_1 \approx t_2)^v = \begin{cases} 1 & \text{if } t_1^v = t_2^v, \\ 0 & \text{otherwise.} \end{cases}$$

[2]–[6] Same as in Definition 7.2.2.

[7] $\exists x A(x)^v = \begin{cases} 1 & \text{if for some } \alpha \in D(v),\ A(u)^{v(u/\alpha)} = 1, \text{ u not} \\ & \text{occurring in } A(x), \\ 0 & \text{otherwise.} \end{cases}$

[8] $\forall x A(x)^v = \begin{cases} 1 & \text{if for every } v' \in K \text{ such that } vRv', \text{ and} \\ & \text{for every } \alpha \in D(v'),\ A(u)^{v'(u/\alpha)} = 1, \text{ u not} \\ & \text{occurring in } A(x), \\ 0 & \text{otherwise.} \end{cases}$

Suppose v is a valuation for $\mathcal{L}$. If $A \notin Form(\mathcal{L})$ (that is, A contains non-logical symbols or free variable symbols not in $\mathcal{L}$), A^v is said to be *undefined*.

Theorem 7.2.7.

Suppose K and R are given, and $v \in K$. For any $A \in Form(\mathcal{L}^p) \cup Form(\mathcal{L})$, $A^v \in \{0,1\}$. If $v, v' \in K$ and vRv', then $A^v = 1$ implies $A^{v'} = 1$. □

Remarks

(1) Theorem 7.2.7 is to be proved by induction on the structure of A. In case A is $\forall xB(x)$ or $\exists xB(x)$, valuations $v(u/\alpha)$ and $v'(u/\alpha)$, which are constructed respectively from v and v', will be used. They are not in K, but we may regard K to be extended to contain them.

(2) Since $v(u/\alpha)$ and $v'(u/\alpha)$ may be different respectively from v and v' only when u is valuated, and

$$u^{v(u/\alpha)} = \alpha = u^{v'(u/\alpha)},$$

we obtain $v(u/\alpha)Rv'(u/\alpha)$ from vRv'.

Definition 7.2.8. (***C-satisfiability, C-validity, C-logical consequence***)

Suppose $\Sigma \subseteq Form(\mathcal{L})$ and $A \in Form(\mathcal{L})$.

Σ is *C-satisfiable* (that is, satisfiable in the constructive sense), iff there are some K, R, and $v \in K$ such that $\Sigma^v = 1$.

A is *C-valid* (that is, valid in the constructive sense), iff for every K, R, and $v \in K$, $A^v = 1$.

$\Sigma \models_C A$ (that is, $\Sigma \models A$ in the constructive sense), iff for every K, R, and $v \in K$, $\Sigma^v = 1$ implies $A^v = 1$.

Of course, Definition 7.2.8 is also available for $\Sigma \subseteq Form(\mathcal{L}^p)$ and $A \in Form(\mathcal{L}^p)$.

Throughout this chapter, K will always be a certain set of valuations and R a reflexive and transitive relation on K.

7.3. FORMAL DEDUCTION

The rules of formal deduction for constructive logic differ from those for classical logic only in that the rule $(\neg-)$ is replaced in constructive logic by the following two weaker rules:

$$
\begin{aligned}
(\neg+)\quad & \text{If } \Sigma, A \vdash B,\\
& \quad \Sigma, A \vdash \neg B,\\
& \text{then } \ \Sigma \vdash \neg A.
\end{aligned}
$$

$$
\begin{aligned}
(\neg)\quad & \text{If } \Sigma \vdash A,\\
& \quad \Sigma \vdash \neg A,\\
& \text{then } \ \Sigma \vdash B.
\end{aligned}
$$

where $(\neg)$ signifies that from contradictory premises any conclusion can be deduced.

Formal deducibility in constructive logic is denoted by the notation $\vdash_C$. Hence $\vdash$ should be replaced by $\vdash_C$ in the rules of formal deduction, in the schemes of formal deducibility, and in the formal proofs for constructive logic. But for convenience, $\vdash$ will be used instead of $\vdash_C$ in the formal proofs for constructive logic.

The definition of $\Sigma \vdash_C A$ will be omitted.

When $\emptyset \vdash_C A$ holds, A is called *C-formally provable* or simply *C-provable*, that is, (formally) provable in the constructive sense.

Σ is *C-consistent* (that is, consistent in the constructive sense) iff there is no A such that $\Sigma \vdash_C A$ and $\Sigma \vdash_C \neg A$.

Σ is *C-maximal consistent* (that is, maximal consistent in the constructive sense) iff Σ is C-consistent and $\Sigma \cup \{A\}$ is not C-consistent (or is C-inconsistent) for any $A \notin \Sigma$.

Corresponding to the classical axiomatic deduction systems described in Chapter 4, the constructive systems can be obtained by replacing the axiom

$$(\neg A \to B) \to ((\neg A \to \neg B) \to A)$$

by two weaker axioms:

$$
\begin{aligned}
&(A \to B) \to ((A \to \neg B) \to \neg A),\\
&\neg A \to (A \to B).
\end{aligned}
$$

Then we can define $\Sigma \vdash_C A$, and prove that for any Σ and A,

$$\Sigma \vdash_C A \text{ iff } \Sigma \vdash_C A.$$

(Refer to Chapter 4.)

Since each of the rules of formal deduction for constructive logic holds in classical logic, we have, for any Σ and A,

1) $$\Sigma \vdash_C A \Longrightarrow \Sigma \vdash A.$$

The converse of 1) does not hold. But we can adopt in constructive logic all those schemes of formal deducibility of classical logic which are established without the aid of $(\neg -)$. We shall list in the following theorem an interesting part of them.

Theorem 7.3.1.

[1] If $A \in \Sigma$, then $\Sigma \vdash_C A$.

[2] If $\Sigma \vdash_C A$, then there is some finite $\Sigma^\circ \subseteq \Sigma$ such that $\Sigma^\circ \vdash_C A$.

[3] If $\Sigma \vdash_C \Sigma'$,
$\Sigma' \vdash_C A$,
then $\Sigma \vdash_C A$.

[4] $A \vdash_C \neg\neg A$.

[5] $A \to B \vdash_C \neg B \to \neg A$.

[6] $A \to B \vdash_C \neg\neg A \to \neg\neg B$.

[7] If $A \vdash_C B$, then $\neg B \vdash_C \neg A$.

[8] If $A \vdash_C B$, then $\neg\neg A \vdash_C \neg\neg B$.

[9] $\emptyset \vdash_C \neg(A \wedge \neg A)$.

[10] $\emptyset \vdash_C \neg\neg(A \vee \neg A)$.

[11] $\emptyset \vdash_C \neg\neg(\neg\neg A \to A)$.

[12] $\neg(A \vee B) \dashv\vdash_C \neg A \wedge \neg B$.

[13] $A \vee B \vdash_C \neg(\neg A \wedge \neg B)$.

[14] $\neg A \vee \neg B \vdash_C \neg(A \wedge B)$.

[15] $A \wedge B \vdash_C \neg(\neg A \vee \neg B)$.

[16] $A \vee B \vdash_C \neg A \to B$.

[17] $\neg A \vee B \vdash_C A \to B$.

[18] $\neg(A \wedge B) \dashv\vdash_C A \to \neg B$.

[19] $A \wedge B \vdash_C \neg(A \to \neg B)$.

[20] $A \to B \vdash_C \neg(A \wedge \neg B)$.

[21] $A \wedge \neg B \vdash_C \neg(A \to B)$.

[22] $\neg\exists x A(x) \dashv\vdash_C \forall x \neg A(x)$.

[23] $\exists x A(x) \vdash_C \neg\forall x \neg A(x)$.

[24] $\forall x A(x) \vdash_C \neg\exists x \neg A(x)$.

[25] $\exists x \neg A(x) \vdash_C \neg\forall x A(x)$.

Theorem 7.3.1 [1] and [3] are still written as (ϵ) and (Tr) respectively. Theorem 7.3.1 [2] is analogous to Theorem 2.6.2. The proof of Theorem 7.3.1 is left to the reader.

It will be pointed out that the following:

$$\neg A \to B \vdash_C \neg B \to A$$
$$\neg A \to \neg B \vdash_C B \to A$$
$$\emptyset \vdash_C A \vee \neg A$$
$$\neg(\neg A \wedge \neg B) \vdash_C A \vee B$$
$$\neg(\neg A \vee \neg B) \vdash_C A \wedge B$$
$$\neg(A \wedge B) \vdash_C \neg A \vee \neg B$$
$$\neg A \to B \vdash_C A \vee B$$
$$A \to B \vdash_C \neg A \vee B$$
$$\neg(A \to \neg B) \vdash_C A \wedge B$$
$$\neg(A \wedge \neg B) \vdash_C A \to B$$
$$\neg(A \to B) \vdash_C A \wedge \neg B$$
$$\neg\forall x \neg A(x) \vdash_C \exists x A(x)$$
$$\neg\exists x \neg A(x) \vdash_C \forall x A(x)$$
$$\neg\forall x A(x) \vdash_C \exists x \neg A(x)$$

do not hold. For a proof of this, refer to the notion of independence in Section 5.6.

Although the converse of 1) does not hold, the formal deducibility in classical logic can be translated into constructive logic in certain ways. These will be formulated in Theorem 7.3.3 and 7.3.7.

Lemma 7.3.2.

[1] $\neg\neg\neg A \dashv\vdash_C \neg A$.
[2] $\neg\neg(A \wedge B) \dashv\vdash_C \neg\neg A \wedge \neg\neg B$.
[3] $\neg\neg(A \to B) \dashv\vdash_C \neg\neg A \to \neg\neg B$.
[4] $\neg\neg(A \leftrightarrow B) \dashv\vdash_C \neg\neg A \leftrightarrow \neg\neg B$.
[5] $\neg\neg\forall x A(x) \vdash_C \forall x \neg\neg A(x)$.

Proof. We shall prove [2] and [5]. The rest are left to the reader.

Proof of [2].

(1) $A \wedge B \vdash A, B$.

(2) $\neg\neg(A \wedge B) \vdash \neg\neg A, \neg\neg B$ (by Thm 7.3.1 [8], (1)).
(3) $\neg\neg(A \wedge B) \vdash \neg\neg A \wedge \neg\neg B$ (by (2)).
(4) $A, B \vdash A \wedge B$.
(5) $\neg\neg A \wedge \neg\neg B, \neg(A \wedge B), A, B \vdash (A \wedge B)$ (by (4)).
(6) $\neg\neg A \wedge \neg\neg B, \neg(A \wedge B), A, B \vdash \neg(A \wedge B)$.
(7) $\neg\neg A \wedge \neg\neg B, \neg(A \wedge B), A \vdash \neg B$ (by (5), (6)).
(8) $\neg\neg A \wedge \neg\neg B, \neg(A \wedge B), A \vdash \neg\neg B$.
(9) $\neg\neg A \wedge \neg\neg B, \neg(A \wedge B) \vdash \neg A$ (by (7), (8)).
(10) $\neg\neg A \wedge \neg\neg B, \neg(A \wedge B) \vdash \neg\neg A$.
(11) $\neg\neg A \wedge \neg\neg B \vdash \neg\neg(A \wedge B)$ (by (9), (10)).
(12) $\neg\neg(A \wedge B) \dashv\vdash \neg\neg A \wedge \neg\neg B$ (by (3), (11)).

Proof of [5].

(1) $\exists x \neg A(x) \vdash \neg\forall x A(x)$ (by Thm 7.3.1 [25]).
(2) $\neg\neg\forall x A(x) \vdash \neg\exists x \neg A(x)$ (by Thm 7.3.1 [5], (1)).
(3) $\neg\exists x \neg A(x) \vdash \forall x \neg\neg A(x)$ (by Thm 7.3.1 [22]).
(4) $\neg\neg\forall x A(x) \vdash \forall x \neg\neg A(x)$ (by (2), (3)). □

Let $\neg\Sigma = \{\neg A | A \in \Sigma\}$. Then we have the following theorem due to Glivenko.

Theorem 7.3.3.
For propositional logic, $\Sigma \vdash A$ iff $\neg\neg\Sigma \vdash_C \neg\neg A$.

Proof. It is obvious that if $\neg\neg\Sigma \vdash_C \neg\neg A$ then $\Sigma \vdash A$. The converse is to be proved by induction on the structure of $\Sigma \vdash A$. We distinguish eleven cases for the rules of formal deduction of classical propositional logic.

For the cases of (Ref) and (+) the converse is obvious. From among the other cases we choose to prove for that of $(\neg -)$. The rest are left to the reader.

Case of $(\neg -)$. We shall prove

$$\begin{aligned} \text{If} \quad & \neg\neg\Sigma, \neg\neg\neg A \vdash_C \neg\neg B, \\ & \neg\neg\Sigma, \neg\neg\neg A \vdash_C \neg\neg\neg B, \\ \text{then} \quad & \neg\neg\Sigma \vdash_C \neg\neg A. \end{aligned}$$

The proof is as follows:

(1) $\neg\neg\Sigma, \neg\neg\neg A \vdash \neg\neg B$ (by supposition).
(2) $\neg\neg\Sigma', \neg\neg\neg A \vdash \neg\neg B$ (by Thm 7.3.1 [3], (1); use finite $\Sigma' \subseteq \Sigma$).

(3) $\neg A \vdash \neg\neg\neg A$ (by Thm 7.3.1 [4]).
(4) $\neg\neg\Sigma, \neg A \vdash \neg\neg\neg A$ (by (3)).
(5) $\neg\neg\Sigma, \neg A \vdash \neg\neg\Sigma'$ (by (ϵ)).
(6) $\neg\neg\Sigma, \neg A \vdash \neg\neg B$ (by (5), (4), (2)).
(7) $\neg\neg\Sigma, \neg A \vdash \neg\neg\neg B$ (analogous to (6)).
(8) $\neg\neg\Sigma \vdash \neg\neg A$ (by ($\neg+$), (6), (7)). □

For first-order logic, a formula A is translated into A°, called the Gödel translation of A. Then let $\Sigma^\circ = \{A^\circ \mid A \in \Sigma\}$.

Definition 7.3.4. (***Gödel translation***)

The *Gödel translation* of formulas of $\mathcal{L}$ is defined by recursion:

[1] $A^\circ = \neg\neg A$ for atom A.
[2] $(\neg A)^\circ = \neg A^\circ$.
[3] $(A \wedge B)^\circ = A^\circ \wedge B^\circ$.
[4] $(A \vee B)^\circ = \neg(\neg A^\circ \wedge \neg B^\circ)$.
[5] $(A \to B)^\circ = A^\circ \to B^\circ$.
[6] $(A \leftrightarrow B)^\circ = A^\circ \leftrightarrow B^\circ$.
[7] $(\forall x A(x))^\circ = \forall x (A(x))^\circ$.
[8] $(\exists x A(x))^\circ = \neg\forall x \neg(A(x))^\circ$.

Lemma 7.3.5.
$A \vdash\dashv A^\circ$. □

Lemma 7.3.6.
$A^\circ \vdash\dashv_C \neg\neg A^\circ$.

Proof. By Theorem 7.3.1 [4] we have $A^\circ \vdash_C \neg\neg A^\circ$. The converse $\neg\neg A^\circ \vdash_C A^\circ$ is to be proved by induction on the structure of A. □

Theorem 7.3.7.
$\Sigma \vdash A$ iff $\Sigma^\circ \vdash_C A^\circ$.

Proof. We first prove $\Sigma^\circ \vdash_C A^\circ \Longrightarrow \Sigma \vdash A$. By $\Sigma^\circ \vdash_C A^\circ$ we have $\Sigma^\circ \vdash A^\circ$. By Theorem 2.6.2, there are $B_1^\circ, \ldots, B_k^\circ \in \Sigma^\circ$ such that

$$B_1^\circ, \ldots, B_k^\circ \vdash A^\circ.$$

By Lemma 7.3.5 we have

$$B_1 \vdash\dashv B_1^\circ,$$
$$\dots$$
$$B_k \vdash\dashv B_k^\circ,$$
$$A \vdash\dashv A^\circ.$$

Then we obtain $B_1, \dots, B_k \vdash A$, and accordingly $\Sigma \vdash A$.

The converse $\Sigma \vdash A \Longrightarrow \Sigma^\circ \vdash_C A^\circ$ will be proved by induction on the structure of $\Sigma \vdash A$. For the rules of formal deduction of non-constructive first-order logic with equality, we distinguish seventeen cases, from among which we shall prove the cases of $(\vee-)$, $(\exists-)$, and $(\approx -)$. The rest are left to the reader.

Case of $(\vee-)$. We shall prove:

$$\begin{aligned} \text{If} \quad & \Sigma^\circ, A^\circ \vdash_C C^\circ, \\ & \Sigma^\circ, B^\circ \vdash_C C^\circ, \\ \text{then} \quad & \Sigma^\circ, (A \vee B)^\circ \vdash_C C^\circ \\ & \quad (\text{that is, } \Sigma^\circ, \neg(\neg A^\circ \wedge \neg B^\circ) \vdash_C C^\circ). \end{aligned}$$

The proof is as follows:

(1) $\Sigma^\circ, A^\circ \vdash C^\circ$
$\Sigma^\circ, B^\circ \vdash C^\circ$ (by supposition).
(2) $\Sigma^\circ, \neg C^\circ \vdash \neg A^\circ \wedge \neg B^\circ$ (by (1)).
(3) $\Sigma^\circ, \neg(\neg A^\circ \wedge \neg B^\circ) \vdash \neg\neg C^\circ$ (by (2)).
(4) $\neg\neg C^\circ \vdash C^\circ$ (by Lem 7.3.6).
(5) $\Sigma^\circ, \neg(\neg A^\circ \wedge \neg B^\circ) \vdash C^\circ$ (by (3), (4)).

Case of $(\exists-)$. We shall prove (writing $A^\circ(u)$ for $(A(u))^\circ$):

$$\begin{aligned} \text{If} \quad & \Sigma^\circ, A^\circ(u) \vdash_C B^\circ, u \text{ not occurring in } \Sigma^\circ \text{ or } B^\circ, \\ \text{then} \quad & \Sigma^\circ, (\exists x A(x))^\circ \vdash_C B^\circ \\ & \quad (\text{that is, } \Sigma^\circ, \neg\forall x \neg A^\circ(x) \vdash_C B^\circ). \end{aligned}$$

The proof is as follows:

(1) $\Sigma^\circ, A^\circ(u) \vdash B^\circ$ (by supposition).
(2) $\Sigma^\circ, \neg B^\circ \vdash \neg A^\circ(u)$ (by (1)).
(3) $\Sigma^\circ, \neg B^\circ \vdash \forall x \neg A^\circ(x)$ (by (2)).
(4) $\Sigma^\circ, \neg\forall x \neg A^\circ(x) \vdash \neg\neg B^\circ$ (by (3)).

(5) $\neg\neg B^\circ \vdash B^\circ$ (by Lem 7.3.6).
(6) $\Sigma^\circ, \neg\forall x\neg A^\circ(x) \vdash B^\circ$ (by (4), (5)).

Case of $(\approx -)$. We shall prove:

$$\begin{array}{ll} \text{If} & \Sigma^\circ \vdash_C A^\circ(t_1), \\ & \Sigma^\circ \vdash_C (t_1 \approx t_2)^\circ \\ & (\text{that is, } \Sigma^\circ \vdash_C \neg\neg(t_1 \approx t_2)), \\ \text{then} & \Sigma^\circ \vdash_C A^\circ(t_2). \end{array}$$

It can be reduced to proving

$$(*) \qquad A^\circ(t_1), \neg\neg(t_1 \approx t_2) \vdash_C A^\circ(t_2)$$

because it can be obtained by $(*)$ and (Tr).

We can prove that $(*)$ is equivalent to

$$(**) \qquad A^\circ(t_2), \neg\neg(t_1 \approx t_2) \vdash_C A^\circ(t_1).$$

The proof of $(**)$ from $(*)$ is as follows:

(1) $A^\circ(t_2), \neg\neg(t_2 \approx t_1) \vdash A^\circ(t_1)$ (by $(*)$).
(2) $t_1 \approx t_2 \vdash t_2 \approx t_1$.
(3) $\neg\neg(t_1 \approx t_2) \vdash \neg\neg(t_2 \approx t_1)$ (by Thm 7.3.1 [8], (2)).
(4) $A^\circ(t_2), \neg\neg(t_1 \approx t_2) \vdash A^\circ(t_2)$.
(5) $A^\circ(t_2), \neg\neg(t_1 \approx t_2) \vdash \neg\neg(t_2 \approx t_1)$ (by (3)).
(6) $A^\circ(t_2), \neg\neg(t_1 \approx t_2) \vdash A^\circ(t_1)$ (by (4), (5), (1)).

Similarly for the proof of $(*)$ from $(**)$. Hence we will prove $(*)$ and $(**)$ simultaneously. The proof is by induction on the structure of $A(t_1)$ and is left to the reader. □

The theorems of replaceability of (both logically and syntactically) equivalent formulas hold in constructive logic as well.

Exercises 7.3.

7.3.1. Prove $(*)$ (simultaneously with $(**)$) as stated in the case of $(\approx -)$ in the proof of Theorem 7.3.7.

7.3.2. Prove for propositional logic $\neg\Sigma \vdash \neg A$ iff $\neg\Sigma \vdash_C \neg A$.

7.3.3. For propositional logic, let A' be defined as:

(1) $A' = \neg\neg A$ for atom A.
(2) $(\neg A)' = \neg A'$.
(3) $(A \wedge B)' = A' \wedge B'$.
(4) $(A \vee B)' = A' \vee B'$.
(5) $(A \to B)' = A' \to B'$.
(6) $(A \leftrightarrow B)' = A' \leftrightarrow B'$.

and let $\Sigma' = \{A' \mid A \in \Sigma\}$. Prove

[1] $\neg\neg A \vdash\dashv_C A'$, A being $\vee$-free.
[2] $\Sigma \vdash A$ iff $\Sigma' \vdash_C A'$, Σ and A being $\vee$-free.

7.4. SOUNDNESS

Theorem 7.4.1. (***Soundness***)
Suppose $\Sigma \subseteq Form(\mathcal{L})$, $A \in Form(\mathcal{L})$.
[1] If $\Sigma \vdash_C A$, then $\Sigma \models_C A$.
[2] if A is C-provable, then A is C-valid.
[3] If Σ is C-satisfiable, then Σ is C-consistent.

Proof. Only [1] needs to be proved. It will be proved by induction on the structure of $\Sigma \vdash_C A$. From among the eighteen cases of the rules of formal deduction of constructive first-order logic with equality, we shall prove [1] for the cases of $(\to +)$, $(\neg +)$, and $(\forall +)$. The rest are left to the reader.

Case of $(\to +)$. We shall prove:

$$\begin{aligned} \text{If} \quad & \Sigma, A \models_C B, \\ \text{then} \quad & \Sigma \models_C A \to B. \end{aligned}$$

Suppose $\Sigma \not\models_C A \to B$, that is, there are some K and R (see Definition 7.2.1), and some $v \in K$ such that $\Sigma^v = 1$ and $(A \to B)^v = 0$. Accordingly, by Definiton 7.2.2, there is some $v' \in K$ such that vRv' and $A^{v'} = 1$, $B^{v'} = 0$. Since vRv', we have $\Sigma^{v'} = 1$ and then $B^{v'} = 1$, thus yielding a contradiction. Hence $\Sigma \models_C A \to B$.

Case of $(\neg +)$. We shall prove:

$$\begin{aligned} \text{If} \quad & \Sigma, A \models_C B, \\ & \Sigma, A \models_C \neg B, \\ \text{then} \quad & \Sigma \models_C \neg A. \end{aligned}$$

Suppose $\Sigma \not\models_C \neg A$, that is, there are some K, R, and $v \in K$ such that $\Sigma^v = 1$ and $(\neg A)^v = 0$. Then, by Definition 7.2.2, there is some $v' \in K$ such that vRv' and $A^{v'} = 1$. Since vRv', we have $\Sigma^{v'} = 1$ and accordingly $B^{v'} = 1$ and $(\neg B)^{v'} = 1$, which is a contradiction. Hence $\Sigma \models_C \neg A$.

Case of ($\forall$+). We shall prove:

$$\text{If} \quad \Sigma \models_C A(u), u \text{ not occurring in } \Sigma,$$
$$\text{then} \quad \Sigma \models_C \forall x A(x).$$

Given any K, R, and $v \in K$. Suppose $\Sigma^v = 1$, v' is any valuation in K such that vRv', and α is any member of $D(v')$. Then $\Sigma^{v'} = 1$. Since u does not occur in Σ, we have $\Sigma^{v'(u/\alpha)} = 1$. Accordingly $A(u)^{v'(u/\alpha)} = 1$ and $\forall x A(x)^v = 1$. Hence $\Sigma \models_C \forall x A(x)$. □

Now we can show that the law of excluded middle does not hold in constructive logic. Suppose $p \vee \neg p$ (p being a proposition symbol) is C-provable. Then, by the Soundness Theorem, it is C-valid.

Let $K = \{v, v'\}$ and R be a reflexive and transitive relation on K such that vRv, vRv', and $v'Rv'$. Besides that, let $p^v = 0$ and $p^{v'} = 1$. Since $p^{v'} = 1$, we have $(\neg p)^v = 0$. Hence $(p \vee \neg p)^v = 0$, contradicting the C-validity of $p \vee \neg p$.

7.5. COMPLETENESS

For simplicity of description we will omit the equality symbol in $\mathcal{L}$. The completeness of constructive first-order logic with equality can be established with the aid of that of the system without equality. This is analogous to the case of classical logic treated in Chapter 5.

Definition 7.5.1. (***Strong consistency***)
$\Sigma \subseteq Form(\mathcal{L})$ is *strong consistent* iff Σ satisfies the following:
[1] Σ is C-consistent.
[2] $\Sigma \vdash_C A$ implies $A \in \Sigma$.
[3] $A \vee B \in \Sigma$ implies $A \in \Sigma$ or $B \in \Sigma$.
[4] $\exists x A(x) \in \Sigma$ implies $A(t) \in \Sigma$ for some $t \in Term(\mathcal{L})$.

In the above definition, [2] is the property in which Σ is closed under C-formal deducibility (refer to Section 5.3), [4] is the existence property (see Defintiion 5.4.1), and [3] is called the *disjunction property.*

Note that strong consistency should not be confused with maximal consistency.

Lemma 7.5.2.

Suppose $\mathcal{L}' = \mathcal{L} \cup \mathcal{D}$, $\mathcal{D}$ being a countable set of new free variable symbols not in $\mathcal{L}$, $\Sigma \subseteq Form(\mathcal{L})$ and $A \in Form(\mathcal{L})$ such that $\Sigma \nvdash_C A$. Then Σ can be extended to some $\Sigma' \subseteq Form(\mathcal{L}')$ such that Σ' is strong consistent and $\Sigma' \nvdash_C A$. ($Form(\Sigma')$ is the set of formulas of $\mathcal{L}'$.)

Proof. $Form(\mathcal{L}')$ is countable. Suppose

$$\mathrm{B}_0, \mathrm{B}_1, \mathrm{B}_2, \ldots \tag{1}$$

is an arbitrary enumeration of it. Construct an infinite sequence of $\Sigma_n \subseteq Form(\mathcal{L}')$ as follows where $n \geq 0$.

Let $\Sigma_0 = \Sigma$. To define Σ_{n+1} from Σ_n we distinguish four cases:

[1] If Σ_n, $B_n \vdash_C A$, set $\Sigma_{n+1} = \Sigma_n$.

[2] If Σ_n, $B_n \nvdash_C A$ and B_n is not a disjunction nor an existential formula, set $\Sigma_{n+1} = \Sigma_n, B_n$.

[3] If Σ_n, $B_n \nvdash_C A$ and B_n is a disjunction $C_1 \vee C_2$, then Σ_n, B_n, $C_1 \nvdash_C A$ or Σ_n, B_n, $C_2 \nvdash_C A$. Set

$$\Sigma_{n+1} = \begin{cases} \Sigma_n, B_n, C_1 & \text{if } \Sigma_n,\ B_n,\ C_1 \nvdash_C A, \\ \Sigma_n, B_n, C_2 & \text{if } \Sigma_n,\ B_n,\ C_2 \nvdash_C A. \end{cases}$$

[4] If Σ_n, $B_n \nvdash_C A$ and B_n is an existential formula $\exists x C(x)$, then, since only finitely many elements of $\mathcal{D}$ occur in Σ_n and B_n, we can find some $d \in \mathcal{D}$ such that d does occur in Σ_n or B_n (of course d does not occur in A) and Σ_n, B_n, $C(d) \nvdash_C A$. Set $\Sigma_{n+1} = \Sigma_n, B_n, C(d)$.

Obviously, we have $\Sigma_n \subseteq \Sigma_{n+1}$ and

$$\Sigma_n \nvdash_C A. \tag{2}$$

Let $\Sigma' = \bigcup_{n \in N} \Sigma_n \subseteq Form(\mathcal{L}')$. Then $\Sigma \subseteq \Sigma'$. We will prove

$$\Sigma' \nvdash_C A \tag{3}$$

and the strong consistency of Σ'. Suppose $\Sigma' \vdash_C A$. Then there are $A_1, \ldots, A_k \in \Sigma'$ such that $A_1, \ldots, A_k \vdash_C A$. Suppose $A_1 \in \Sigma_{i_1}, \ldots, A_k \in$

Σ_{i_k} and $i = max(i_1, \ldots, i_k)$. Then $A_1, \ldots, A_k \in \Sigma_i$ and $\Sigma_i \vdash_C A$, contradicting (2). Hence we have (3). Accordingly, Σ' is C-consistent and satisfies the condition [1] in Definition 7.5.1.

Suppose $C \in Form(\mathcal{L}')$ and $\Sigma' \vdash_C C$. Then we have

$$\Sigma', C \nvdash_C A. \tag{4}$$

(If $\Sigma', C \vdash_C A$, we have $\Sigma' \vdash_C C \to A$ and then $\Sigma' \vdash_C A$, thus contradicting (3).) Suppose C is B_m in the enumeration (1). Then $\Sigma_m, B_m \nvdash_C A$. (If $\Sigma_m, B_m \vdash_C A$, we have $\Sigma', C \vdash_C A$ because $\Sigma_m \subseteq \Sigma'$, contradicting (4).) By [2]–[4], we have $B_m \in \Sigma_{m+1}$ and accordingly $C \in \Sigma'$. Hence Σ' satisfies [2] of Definition 7.5.1.

Suppose $C_1 \vee C_2 \in \Sigma'$ and $C_1 \vee C_2$ is B_m in (1). Then $\Sigma_m, C_1 \vee C_2 \nvdash_C A$. (If $\Sigma_m, C_1 \vee C_2 \vdash_C A$, then $\Sigma', C_1 \vee C_2 \vdash_C A$ and hence $\Sigma' \vdash_C A$, contradicting (3).) By [3] we have $C_1 \in \Sigma_{m+1}$ or $C_2 \in \Sigma_{m+1}$. Hence $C_1 \in \Sigma'$ or $C_2 \in \Sigma'$, and Σ' satisfies [3] of Definition 7.5.1.

Suppose $\exists x C(x) \in \Sigma'$ and $\exists x C(x)$ is B_m in (1). Then $\Sigma_m, \exists x C(x) \nvdash_C A$. By [4] we have $C(d) \in \Sigma_{m+1}$ for some $d \in \mathcal{D}$ and $C(d) \in \Sigma'$. Hence Σ' satisfies [4] of Definition 7.5.1, and Σ' is strong consistent. □

Suppose $\mathcal{D}_0, \mathcal{D}_1, \mathcal{D}_2, \ldots$ are countable sets of new free variable symbols and $\mathcal{L}, \mathcal{D}_0, \mathcal{D}_1, \mathcal{D}_2, \ldots$ are pairwise disjoint.

Let

$$\begin{aligned} &\mathcal{L}_0 = \mathcal{L}, \\ &\mathcal{L}_{n+1} = \mathcal{L}_n \cup \mathcal{D}_n (n \geq 0), \\ &\mathcal{D} = \bigcup_{n \in N} \mathcal{D}_n, \\ &\mathcal{L}' = \mathcal{L} \cup \mathcal{D}. \end{aligned}$$

Then $Term(\mathcal{L}_n)$ and $Term(\mathcal{L}')$ are respectively, sets of terms of $\mathcal{L}_n$ and $\mathcal{L}'$; $Form(\mathcal{L}_n)$ and $Form(\mathcal{L}')$ are respectively, sets of formulas of $\mathcal{L}_n$ and $\mathcal{L}'$.

Suppose $\Sigma \subseteq Form(\mathcal{L})$, $A \in Form(\mathcal{L})$, and $\Sigma \nvdash_C A$. Let $\Sigma_0 = \Sigma$. By Lemma 7.5.2, Σ_0 can be extended to some $\Sigma_1 \subseteq Form(\mathcal{L}_1)$ such that Σ_1 is strong consistent and $\Sigma_1 \nvdash_C A$. Similarly, Σ_1 can be extended to some $\Sigma_2 \subseteq Form(\mathcal{L}_2)$ such that Σ_2 is strong consistent and $\Sigma_2 \nvdash_C A$, etc. Therefore, for $n \geq 1$, we have $\Sigma_n \subseteq Form(\mathcal{L}_n)$ such that Σ_n is strong consistent, $\Sigma_n \nvdash_C A$, and $\Sigma_n \subseteq \Sigma_{n+1}$.

For $\mathcal{L}_n$ $(n \geq 0)$ we construct a valuation v_n as follows. Let

$$D(v_n) = \{t' \mid t \in Term(\mathcal{L}_n)\}$$

be the domain of v_n.

For any individual symbol a, let $a^{v_n} = a'$.

For any free variable symbol u in $\mathcal{L}_n$, let $u^{v_n} = u'$.

For any k-ary function symbol f and any $t'_1, \ldots, t'_k \in D(v_n)$, let $f^{v_n}(t'_1, \ldots, t'_k) = f(t_1, \ldots, t_k)'$. Then $t^{v_n} = t' \in D(v_n)$ for any $t \in Term(\mathcal{L}_n)$, and $t^{v_n} = t^{v_s}$ for any $s \geq n$.

For any k-ary relation symbol F and any $t', \ldots, t'_k \in D(v_n)$, let $\langle t'_1, \ldots, t'_k \rangle \in F^{v_n}$ iff $F(t_1, \ldots, t_k) \in \Sigma_n$. That is, for any $t_1, \ldots, t_k \in Term\ (\mathcal{L}_n)$, $F(t_1, \ldots, t_k)^{v_n} = 1$ iff $F(t_1, \ldots, t_k) \in \Sigma_n$.

Let $K = \{v_0, v_1, v_2, \ldots\}$ and R be a binary relation on K such that $v_i R v_j$ iff $i \leq j$. Then R is reflexive and transitive. Suppose $v_i R v_j$. Obviously we have $D(v_i) \subseteq D(v_j)$, $F^{v_i} \subseteq F^{v_j}$ and $f^{v_i} = f^{v_j} | D(v_i)$. Hence K (with its elements) and R satisfy the conditions in Definition 7.2.3.

The conventions described above will be used throughout this section.

Lemma 7.5.3.
Suppose $A \in Form(\mathcal{L}')$. For $n \geq 1$, $A^{v_n} = 1$ iff $A \in \Sigma_n$.

Proof. By induction on the structure of A. We distinguish eight cases, from among which we shall choose to deal in detail with the cases of $A = B \to C$, $\neg B$, $\exists x B(x)$, and $\forall x B(x)$. The rest are left to the reader.

Case of $A = B \to C$. We first prove: $B \to C \in \Sigma_n \Longrightarrow (B \to C)^{v_n} = 1$. Suppose $B \to C \in \Sigma_n$. Take any v_s such that $n \leq s$. Then $\Sigma_n \subseteq \Sigma_s$. Suppose $B^{v_s} = 1$. We have

$$B \to C \in \Sigma_s,$$

$$\Sigma_s \vdash_C B \to C,$$

$$B \in \Sigma_s \quad (\text{by } B^{v_s} = 1, \text{ ind hyp}),$$

$$\Sigma_s \vdash_C B,$$

$$\Sigma_s \vdash_C C,$$

$$C \in \Sigma_s \quad (\text{by strong consis of } \Sigma_s),$$

$$C^{v_s} = 1 \quad (\text{by ind hyp}).$$

Since $v_n R v_s$, we have $(B \to C)^{v_n} = 1$.

Then we prove: $(B \to C)^{v_n} = 1 \Longrightarrow B \to C \in \Sigma_n$. Suppose $(B \to C)^{v_n} = 1$, then $(B \to C)^{v_n}$ is not undefined and accordingly $B \to C \in Form(\mathcal{L}_n)$. Suppose $B \to C \notin \Sigma_n$. Then we have

$$\Sigma_n \nvdash_C B \to C \quad (\text{by strong consis of } \Sigma_n),$$

$$\Sigma_n, B \nvdash_C C.$$

By Lemma 7.5.2, Σ_n, B can be extended to some $\Sigma_{n+1} \subseteq Form(\mathcal{L}_{n+1})$ such that Σ_{n+1} is strong consistent and $\Sigma_{n+1} \nvdash_C$ C. Then

$$\begin{aligned} &C \notin \Sigma_{n+1}, \\ &C^{v_{n+1}} = 0 \quad \text{(by ind hyp and } (B \to C)^{v_n} \text{ not undefined)}, \\ &B^{v_{n+1}} = 1 \quad \text{(by ind hyp and } B \in \Sigma_{n+1}). \end{aligned}$$

Since $v_n R v_{n+1}$, we have $(B \to C)^{v_n} = 0$, contradicting the supposition.

Case of $A = \neg B$. We first prove: $\neg B \in \Sigma_n \Longrightarrow (\neg B)^{v_n} = 1$. Suppose $\neg B \in \Sigma_n$. Take any v_s such that $n \leq s$. We have

$$\begin{aligned} &\neg B \in \Sigma_s, \\ &B \notin \Sigma_s \quad \text{(by strong consis of } \Sigma_s), \\ &B^{v_s} = 0 \quad \text{(by ind hyp and } B^{v_s} \text{ not undefined)}. \end{aligned}$$

Since $v_n R v_s$, we have $(\neg B)^{v_n} = 1$.

We then prove: $(\neg B)^{v_n} = 1 \Longrightarrow \neg B \in \Sigma_n$. Suppose $(\neg B)^{v_n} = 1$, then $(\neg B)^{v_n}$ is not undefined and accordingly $\neg B \in Form(\mathcal{L}_n)$. Suppose $\neg B \notin \Sigma_n$. We have

$$\begin{aligned} &\Sigma_n \nvdash_C \neg B \quad \text{(by strong consis of } \Sigma_n), \\ &\Sigma_n \nvdash_C B \to \neg B \quad (\text{by } B \to \neg B \vdash_C \neg B), \\ &\Sigma_n, B \nvdash_C \neg B. \end{aligned}$$

By Lemma 7.5.2, Σ_n, B can be extended to some $\Sigma_{n+1} \subseteq Form(\mathcal{L}_{n+1})$ such that Σ_{n+1} is strong consistent (and $\Sigma_{n+1} \nvdash_C \neg B$, which is not to be used). Then $B \in \Sigma_{n+1}$ and, by the induction hypothesis, $B^{v_{n+1}} = 1$. Since $v_n R v_{n+1}$, we have $(\neg B)^{v_n} = 0$, contradicting the supposition.

Case of $A = \exists x B(x)$. We first prove: $\exists x B(x) \in \Sigma_n \Longrightarrow \exists x B(x)^{v_n} = 1$. Suppose $\exists x B(x) \in \Sigma_n$. We have

$$\begin{aligned} &B(t) \in \Sigma_n \quad \text{for some } t \in Term(\mathcal{L}_n) \text{ (by strong consis of } \Sigma_n), \\ &B(t)^{v_n} = 1 \quad \text{(by ind hyp)}. \end{aligned}$$

Form B(u), u being a free valuable symbol of $\mathcal{L}_n$ not occurring in B(x) or t, we have $B(u)^{v_n(u/t^{v_n})} = B(t)^{v_n} = 1$, where $t^{v_n} \in D(v_n)$. Hence $\exists x B(x)^{v_n} = 1$.

We then prove: $\exists x B(x)^{v_n} = 1 \Longrightarrow \exists x B(x) \in \Sigma_n$. Suppose $\exists x B(x)^{v_n} = 1$, then $\exists x B(x)^{v_n}$ is not undefined and accordingly $\exists x B(x) \in Form(\mathcal{L}_n)$.

Suppose $\exists x B(x) \notin \Sigma_n$. Then $\Sigma_n \nvdash_C \exists x B(x)$ by the strong consistency of Σ_n. Take any $t' \in D(v_n)$, that is, $t \in Term(\mathcal{L}_n)$. We have

$$\Sigma_n \nvdash_C B(t),$$
$$B(t) \notin \Sigma_n,$$
$$B(t)^{v_n} = 0 \quad \text{(by ind hyp and } B(t)^{v_n} \text{ not undefined).}$$

Form B(u), u being free variable symbol of $\mathcal{L}_n$ not occurring in B(x) or t. Since $t^{v_n} = t'$, we have $B(u)^{v_n(u/t')} = B(u)^{v_n(u/t^{v_n})} = B(t)^{v_n} = 0$. Hence $\exists x B(x)^{v_n} = 0$, contradicting the supposition.

Case of $A = \forall x B(x)$. We first prove: $\forall x B(x) \in \Sigma_n \Longrightarrow \forall x B(x)^{v_n} = 1$. Suppose $\forall x B(x) \in \Sigma_n$. Take any v_s such that $n \leq s$. Since $\Sigma_n \subseteq \Sigma_s$, we have $\forall x B(x) \in \Sigma_s$ and $\Sigma_s \vdash_C \forall x B(x)$. Take any $t' \in D(v_s)$, that is, $t \in Term(\mathcal{L}_s)$. We have

$$\Sigma_s \vdash_C B(t),$$
$$B(t) \in \Sigma_s \quad \text{(by strong consis of } \Sigma_s\text{),}$$
$$B(t)^{v_s} = 1 \quad \text{(by ind hyp and } B(t)^{v_s} \text{ not undefined).}$$

Form B(u), u being free variable symbol of $\mathcal{L}_s$ not occurring in B(x) or t. Since $t^{v_s} = t'$, we have $B(u)^{v_s(u/t')} = B(u)^{v_s(u/t^{v_s})} = B(t)^{v_s} = 1$. Since $v_n R v_s$, we have $\forall x B(x)^{v_n} = 1$.

Then we prove: $\forall x B(x)^{v_n} = 1 \Longrightarrow \forall x B(x) \in \Sigma_n$. Suppose $\forall x B(x)^{v_n} = 1$, then $\forall x B(x)^{v_n}$ is not undefined and accordingly $\forall x B(x) \in Form(\mathcal{L}_n)$. Suppose $\forall x B(x) \notin \Sigma_n$. We have

$$\text{(1)} \qquad \Sigma_n \nvdash_C \forall x B(x)$$

by the strong consistency of Σ_n. Take some $d \in \mathcal{D}_n$. Note that d does not occur in Σ_n or in $\forall x B(x)$ (because $\Sigma_n \subseteq Form(\mathcal{L}_n)$ and $\forall x B(x) \in Form(\mathcal{L}_n)$), and that $B(d) \in Form(\mathcal{L}_{n+1})$ and $B(d) \notin Form(\mathcal{L}_n)$. Then we have

$$\text{(2)} \qquad \Sigma_n \nvdash_C B(d).$$

(If $\Sigma_n \vdash_C B(d)$, then $\Sigma_n \vdash_C \forall x B(x)$, contradicting (1).) We may regard $\Sigma_n \subseteq Form(\mathcal{L}_{n+1})$ and $\forall x B(x) \in Form(\mathcal{L}_{n+1})$. By (2) and Lemma 7.5.2,

Σ_n can be extended to some $\Sigma_{n+2} \subseteq Form(\mathcal{L}_{n+2})$ such that Σ_{n+2} is strong consistent and $\Sigma_{n+2} \nvdash_C B(d)$. Then we obtain

$$B(d) \notin \Sigma_{n+2},$$
$$B(d)^{v_{n+2}} = 0 \quad \text{(by ind hyp and } B(d)^{v_{n+2}} \text{ not undefined)},$$
$$B(d)^{v_{n+2}(d/d^{v_{n+2}})} = B(d)^{v_{n+2}} = 0,$$

where $d^{v_{n+2}} \in D(v_{n+2})$. Since $v_n R v_{n+2}$, we have $\forall x B(x)^{v_n} = 0$, contradicting the supposition. □

Theorem 7.5.4. (***Completeness***)
Suppose $\Sigma \subseteq Form(\mathcal{L})$ and $A \in Form(\mathcal{L})$.
[1] If Σ is C-consistent, then Σ is C-satisfiable.
[2] If $\Sigma \models_C A$, then $\Sigma \vdash_C A$.
[3] If A is C-valid, then A is C-provable.

Proof. We need to prove [1] and [2] only. Suppose Σ is C-consistent. Then there is some $A \in Form(\mathcal{L})$ such that $\Sigma \nvdash_C A$. By Lemma 7.5.2, Σ can be extended to some $\Sigma_1 \subseteq Form(\mathcal{L}_1)$ such that Σ_1 is strong consistent and $\Sigma_1 \nvdash_C A$. Take any $B \in \Sigma$. We have $B \in \Sigma_1$. By Lemma 7.5.3, $B^{v_1} = 1$. Hence $\Sigma^{v_1} = 1$ and Σ is C-satisfiable. Then [1] is proved.

Suppose $\Sigma \nvdash_C A$. We have proved above $\Sigma_1 \nvdash_C A$ and $\Sigma^{v_1} = 1$. By $\Sigma_1 \nvdash_C A$ we have $A \notin \Sigma_1$, and then $A^{v_1} = 0$, by Lemma 7.5.3. Since $\Sigma^{v_1} = 1$ and $A^{v_1} = 0$, we have $\Sigma \not\models_C A$, and [2] is proved. □

8

MODAL PROPOSITIONAL LOGIC

Modal logic is the logic of modal notions. Given any proposition $\mathcal{A}$, we can form the propositions " $\mathcal{A}$ is necessary" and " $\mathcal{A}$ is possible". Necessity and possibility are modal notions. Propositions with and without modal notions are of different kinds, hence modal and non-modal logics are different as well. We shall study modal propositional logic in the present chapter and modal first-order logic in the next one.

8.1. MODAL PROPOSITIONAL LANGUAGE

In classical logic, propositions are either true or false. But in modal logic we shall, among true propositions, distinguish propositions which are necessarily true from those which are not, and among false propositions, distinguish necessarily false propositions from those which are not.

Necessarily true propositions are said to be *necessary*, and necessarily false propositions are said to be *impossible*. Propositions which are not impossible are said to be *possible*. Hence possible propositions include all true propositions (necessary or not).

Necessity and *possibility* are basic modal notions.

Given any proposition $\mathcal{A}$, we can form the proposition "$\mathcal{A}$ is necessary", which means that it is necessary to have $\mathcal{A}$. This proposition will be true when $\mathcal{A}$ is necessary, and false when $\mathcal{A}$ is not. Necessity is a unary modal operator which can be applied on a proposition to form a new one. In this sense it seems to be like negation, but, unlike negation, it is not truth-functional. Although from the falsehood of $\mathcal{A}$ we can assert the falsehood of "$\mathcal{A}$ is necessary", yet from the truth of $\mathcal{A}$ we can assert neither the truth nor the falsehood of "$\mathcal{A}$ is necessary".

Similarly for "$\mathcal{A}$ is possible", which means that it is possible to have $\mathcal{A}$. This proposition is formed by the unary modal operator possibility. From the truth of $\mathcal{A}$ we can assert the truth of "$\mathcal{A}$ is possible", but from the falsehood of $\mathcal{A}$ we can assert neither the truth nor the falsehood of "$\mathcal{A}$ is possible".

Modal logic is the logic of necessity and possibility. Modal logic is also classified into classical and constructive systems. Since classical model logic receives more attention in the literature, we shall confine ourselves in this book to the discussion of such modal systems only.

We shall use the roman-type capital Latin letters

$$\mathrm{L} \quad \mathrm{M}$$

for the necessity and possibility symbols respectively. For simplicity of description, we will use L as the primitive symbol and introduce M by definition. Then the *modal propositional language* $\mathcal{L}^{pm}$ is obtained by adding L to the propositional language $\mathcal{L}^{p}$.

The set $Atom(\mathcal{L}^{pm})$ of atoms of $\mathcal{L}^{pm}$ is the same as $Atom(\mathcal{L}^{p})$. The set $Form(\mathcal{L}^{pm})$ of formulas of $\mathcal{L}^{pm}$ is the smallest set of expressions of $\mathcal{L}^{pm}$ closed under the following formation rules of formulas of $\mathcal{L}^{pm}$:

[1] $Atom(\mathcal{L}^{pm}) \subseteq Form(\mathcal{L}^{pm})$.

[2] If $\mathrm{A} \in Form(\mathcal{L}^{pm})$, then $(\neg\mathrm{A})$, $(\mathrm{LA}) \in Form(\mathcal{L}^{pm})$.

[3] If A, $\mathrm{B} \in Form(\mathcal{L}^{pm})$, then $(\mathrm{A} * \mathrm{B}) \in Form(\mathcal{L}^{pm})$, $*$ being any one of $\wedge$, $\vee$, $\rightarrow$, and $\leftrightarrow$.

The details of the structure of formulas of $\mathcal{L}^{\mathrm{pm}}$ are left to the reader.

8.2. SEMANTICS

Usually the term "world" is used in the discussion of semantics of modal logic. A world is a conceivable state of affairs. According to Chang and

Keisler [1973], "world" is synonymous with "interpretation". As explained in Section 7.2, we have used "valuation" for interpretation of both the propositional and first-order languages. Therefore, we will use "valuation" instead of "world" in the study of modal logic.

As in the case of constructive logic, the semantics of modal logic is established after formal deduction. The formal deduction systems of modal logic are obtained by adding rules concerning necessity and possibility symbols to classical logic.

Since $\mathcal{L}^{pm}$ contains one more symbol L than $\mathcal{L}^{p}$, the valuation for $\mathcal{L}^{pm}$ will be constructed by adding the valuation for L to that for $\mathcal{L}^{p}$. We begin with some intuitive explanations.

Suppose a proposition $\mathcal{A}$ is expressed by a formula A. Then LA expresses "$\mathcal{A}$ is necessary". According to a familiar and natural idea which is often credited to Leibniz, a necessary proposition is one which is true not only in a certain designated valuation, but in all other possible valuations as well. Suppose v is an arbitrary valuation. Then $(\mathrm{LA})^v = 1$ iff for every valuation v', $\mathrm{A}^{v'} = 1$.

For instance, we consider the formulas p and $\mathrm{p} \to \mathrm{p}$, where p is any proposition symbol. Then we have $(\mathrm{Lp})^v = 0$ because there is some v' such that $\mathrm{p}^{v'} = 0$. But $(\mathrm{L}(\mathrm{p} \to \mathrm{p}))^v = 1$ because for every v', $(\mathrm{p} \to \mathrm{p})^{v'} = 1$.

Therefore the truth value of LA is not determined by a certain designated valuation v, but by all valuations including v. The whole set of valuations may be regarded as all those in a certain collection K of valuations.

Then we have the following definitions.

Definition 8.2.1. (***Valuation, value of formulas***)

Suppose K is a set. Each element of K is called a *valuation* for $\mathcal{L}^{pm}$, which is a function with the set of all proposition symbols as domain and $\{1, 0\}$ as range.

The *value* of formulas under valuation $v \in K$ is defined by recursion:

[1] $\mathrm{p}^v \in \{1, 0\}$ for atom p.

[2] $$(\neg \mathrm{A})^v = \begin{cases} 1 & \text{if } \mathrm{A}^v = 0, \\ 0 & \text{otherwise.} \end{cases}$$

[3] $$(\mathrm{A} \wedge \mathrm{B})^v = \begin{cases} 1 & \text{if } \mathrm{A}^v = \mathrm{B}^v = 1, \\ 0 & \text{otherwise.} \end{cases}$$

[4] $(A \vee B)^v = \begin{cases} 1 & \text{if } A^v = 1 \text{ or} B^v = 1, \\ 0 & \text{otherwise.} \end{cases}$

[5] $(A \rightarrow B)^v = \begin{cases} 1 & \text{if } A^v = 0 \text{ or} B^v = 1, \\ 0 & \text{otherwise.} \end{cases}$

[6] $(A \leftrightarrow B)^v = \begin{cases} 1 & \text{if } A^v = B^v, \\ 0 & \text{otherwise.} \end{cases}$

[7] $(LA)^v = \begin{cases} 1 & \text{if for every } v' \in K,\ A^{v'} = 1, \\ 0 & \text{otherwise.} \end{cases}$

Definition 8.2.2. (***Satisfiability, validity***)

Suppose $\Sigma \subseteq Form(\mathcal{L}^{pm})$, $A \in Form(\mathcal{L}^{pm})$.

Σ is *satisfiable* iff there are some set K of valuations and some $v \in K$ such that $\Sigma^v = 1$ in the sense of Definition 8.2.1.

A is *valid* iff for every set K of valuations and every $v \in K$, $A^v = 1$ in the sense of Definition 8.2.1.

We will define valuation and value of formulas in another form.

Definition 8.2.3. (***Valuation, value of formulas***)

Suppose K is a set and R is an equivalence relation on K. Each element of K is called a *valuation* for $\mathcal{L}^{pm}$ as in Definition 8.2.1.

The *value* of formulas under valuation $v \in K$ is defined by recursion:

[1]–[6] Same as in Definition 8.2.1.

[7] $(LA)^v = \begin{cases} 1 & \text{if for every } v' \in K \text{ such that } vRv',\ A^{v'} = 1, \\ 0 & \text{otherwise.} \end{cases}$

Then we formulate the following definition which is equivalent to Definition 8.2.2.

Definition 8.2.4. (***Satisfiability, validity***)

Suppose $\Sigma \subseteq Form(\mathcal{L}^{pm})$, $A \in Form(\mathcal{L}^{pm})$.

Σ is *satisfiable* iff there are some set K of valuations, some equivalence relation R on K, and some $v \in K$ such that $\Sigma^v = 1$ in the sense of Definition 8.2.3.

A is *valid* iff for every set K of valuations, every equivalence relation R on K, and every $v \in K$, $A^v = 1$ in the sense of Definition 8.2.3.

Theorem 8.2.5.
Suppose $\Sigma \subseteq Form(\mathcal{L}^{pm})$ and $\mathrm{A} \in Form(\mathcal{L}^{pm})$.
[1] Σ is satisfiable in the sense of Definition 8.2.2 iff Σ is satisfiable in the sense of Definition 8.2.4.
[2] A is valid in the sense of Definition 8.2.2 iff A is valid in the sense of Definition 8.2.4.

Proof. We shall first prove [1]. Suppose Σ is satisfiable in the sense of Definition 8.2.2. Then we have some set K of valuations and some $v \in K$ such that $\Sigma^v = 1$ in the sense of Definition 8.2.1. Let R be a binary relation on K such that for any v, $v' \in K$, vRv'. Then R is an equivalence relation on K. It is easy to prove that $\Sigma^v = 1$ in the sense of Definition 8.2.3. Hence Σ is satisfiable in the sense of Definition 8.2.4.

For the converse, suppose Σ is satisfiable in the sense of Definition 8.2.4. Then we have some K, some equivalence relation R on K, and some $v \in K$ such that $\Sigma^v = 1$ in the sense of Definition 8.2.3. Let $K' = \{v' | \ vRv'\}$. Then $v \in K'$. By K' and v, we have $\Sigma^v = 1$ in the sense of Definition 8.2.1. Hence Σ is satisfiable in the sense of Definition 8.2.2.

[2] can be proved in a similar way. □

The distinction between the two equivalent definitions is that Definition 8.2.4 is concerned with an equivalence relation R on K, while Definition 8.2.2 is not.

The system of modal propositional logic corresponding with the semantics formulated in the above definitions is called S_5. The valuation, satisfiability, and validity in S_5 are called *S_5-valuation*, *S_5-satisfiability*, and *S_5-validity*.

Other systems of modal propositional logic can be constructed by modifying the requirements of the relation R on K. For instance, R may be merely reflexive, or reflexive and transitive, or reflexive and symmetric. The various systems of modal propositional logic thus obtained are listed as follows:

Modal propositional logic	Requirements of R
T	reflexive
S_4	reflexive and transitive
B	reflexive and symmetric
S_5	equivalence relation

Note that B is here a system of modal propositional logic due to Brouwer. The reader is refered to Hughes and Cresswell [1968] for the names and historical notes of the systems of modal logic.

Then we have the notions of T (S_4, B)-*valuation*, T (S_4, B)-*satisfiability*, and T (S_4, B)-*validity*. Their definitions are left to the reader.

Obviously we have the following statements:

$$\text{A is T-valid} \Longrightarrow \left\{ \begin{matrix} \text{A is } S_4\text{-valid} \\ \text{A is B-valid} \end{matrix} \right\}$$
$$\Longrightarrow \text{A is } S_5\text{-valid.}$$

$$\Sigma \text{ is } S_5\text{-satisfiable} \Longrightarrow \left\{ \begin{matrix} \Sigma \text{ is } S_4\text{-satisfiable} \\ \Sigma \text{ is B-satisfiable} \end{matrix} \right\}$$
$$\Longrightarrow \Sigma \text{ is T-satisfiable.}$$

But S_4-validity (S_4-satisfiability) does not imply, nor is it implied by, B-validity (B-satisfiability).

Logical consequence in these modal systems is defined as in the classical systems (in Chapters 2 and 3) with suitable modifications. We use the notations $\models_T$, $\models_{S_4}$, $\models_B$, $\models_{S_5}$ for them. For instance, $\Sigma \models_{S_5} A$ is defined as follows:

$$\Sigma \models_{S_5} A \quad \text{iff for every set } K \text{ of } S_5\text{-valuations and every } v \in K,\ \Sigma^v = 1 \Longrightarrow A^v = 1 \text{ in the sense of Definition 8.2.1;}$$

or

$$\Sigma \models_{S_5} A \quad \text{iff for every set } K \text{ of } S_5\text{-valuations, every equivalence relation } R \text{ on } K, \text{ and every } v \in K, \Sigma^v = 1 \Longrightarrow A^v = 1 \text{ in the sense of Definition 8.2.3.}$$

$\Sigma \models_T A$, $\Sigma \models_{S_4} A$, and $\Sigma \models_B A$ are defined in a similar way with modifications of the requirements of R. Then we have

$$\Sigma \models_T A \Longrightarrow \left\{ \begin{matrix} \Sigma \models_{S_4} A \\ \Sigma \models_B A \end{matrix} \right\} \Longrightarrow \Sigma \models_{S_5} A.$$

Hence S_5 is a stronger system than both S_4 and B, and both S_4 and B are stronger than T. L in S_5 is a stronger notion than that in S_4 and B, and L in S_4 and B is stronger than that in T. But S_4 is not stronger than B, nor is B stronger than S_4.

Theorem 8.2.6. (***Replaceability of equivalent formulas***)
Suppose B $\models\!\!\!\dashv$ C and A′ results from A by replacing some (not necessarily all) occurrences of B in A by C. Then A $\models\!\!\!\dashv$ A′.

Proof. By induction on the structure of A. The crucial step is to prove, for the case of $A = LA_1$,

$$A_1 \models\!\!\!\dashv A_1' \Longrightarrow LA_1 \models\!\!\!\dashv LA_1'.$$

The proof is left to the reader. □

Theorem 8.2.6 holds in each of the systems T, S_4, B, and S_5.

Of the modal systems mentioned above, the semantics for S_5 seems to be the most natural, because $(LA)^v = 1$ signifies in S_5 that $A^{v'} = 1$ for every $v' \in K$. But S_4 has important applications in temporal logic. (See Manna [1982].)

The semantics presented in the foregoing is due to Kripke. For the relations between modal and constructive logic, Kripke [1965] stated that the semantics for modal logic, together with the known mappings of constructive logic into the modal system S_4, inspired the semantics for constructive logic.

8.3. FORMAL DEDUCTION

To define formal deducibility for the modal systems, we need to introduce the rules of formal deduction for them. Each of these modal systems contains all the eleven rules of non-modal propositional logic and some additional rules concerning the modal symbols.

Firstly, another modal symbol M will be introduced. The formula

$$(MA)$$

which is formed by means of M is defined to be $(\neg(L(\neg A)))$. The formulas (MA) and $(\neg(L(\neg A)))$ may be abbreviated as MA and $\neg L \neg A$. The six rules concerning L and M are formulated as follows:

(L−) If $\Sigma \vdash$ LA,
then $\Sigma \vdash$ A. (L–*elimination*)

(→−(L)) If $\Sigma \vdash$ L(A → B),
$\Sigma \vdash$ LA,
then $\Sigma \vdash$ LB. (→–*elimination in the scope of* L)

(L+) If $\emptyset \vdash$ A,
then $\emptyset \vdash$ LA. (L–*introduction*)

(L+L) If $\Sigma \vdash$ LA,
then $\Sigma \vdash$ LLA. (L–*introduction to* L)

(L+M) If $\Sigma \vdash$ MA,
then $\Sigma \vdash$ LMA. (L–*introduction to* M)

(LM+) If $\Sigma \vdash$ A,
then $\Sigma \vdash$ LMA. (LM–*introduction*)

It should be pointed out that (→−(L)) is distinct from (→−), and that the three L–introduction rules (L+), (L+L), and (L+M) are distinct from each other. Besides, the premise in (L+) must be empty.

Now we are in a position to give the rules of formal deduction contained in these modal systems. First, the weakest system T contains the three rules (L−), (→ −(L)), and (L+) in addition to the eleven rules of non-modal propositional logic. Then S_4, S_5, and B contain, in addition to the rules of T, (L+L), (L+M), and (LM+) respectively.

The notations $\vdash_T$, $\vdash_{S_4}$, $\vdash_{S_5}$, and $\vdash_B$ are used for formal deducibility in these modal systems. We will assert in advance that both (L+L) and (LM+) hold in S_5, but that (L+L) does not hold in B, nor does (LM+) hold in S_4. Hence we have

$$\Sigma \vdash_T A \Longrightarrow \left\{ \begin{matrix} \Sigma \vdash_{S_4} A \\ \Sigma \vdash_B A \end{matrix} \right\} \Longrightarrow \Sigma \vdash_{S_5} A.$$

But $\Sigma \vdash_{S_4} A$ does not imply and is not implied by $\Sigma \vdash_B A$.

The definitions of $\Sigma \vdash_T A$, $\Sigma \vdash_{S_4} A$, $\Sigma \vdash_{S_5} A$, and $\Sigma \vdash_B A$, and those of $T(S_4, S_5, B)$-*formal provability*, $T(S_4, S_5, B)$-*consistency*, and $T(S_4, S_5, B)$-*maximal consistency* are left to the reader.

In the following theorems concerning formal deducibility, we will write $\vdash_T$, $\vdash_{S_4}$, $\vdash_{S_5}$, or $\vdash_B$ to indicate the system in which the theorems hold. But we may omit "T" "S_4", "S_5", and "B" in the proofs if no confusion will arise.

Theorem 2.6.2 also holds in modal logics.

Theorem 8.3.1.

[1] If $A \vdash_T B$,
then $LA \vdash_T LB$.

[2] If $A \dashv\vdash_T B$,
then $LA \dashv\vdash_T LB$.

[3] If $A_1, \ldots, A_n \vdash_T A$,
then $LA_1, \ldots, LA_n \vdash_T LA$.

[4] $A \vdash_T MA$.

[5] $L(A \to B), L(B \to A) \vdash_T LA \leftrightarrow LB$.

[6] $L(A \wedge B) \dashv\vdash_T LA \wedge LB$.

[7] $L(A \leftrightarrow B) \dashv\vdash_T L(A \to B), L(B \to A)$.

Proof. We shall prove [1], [4], and [6]. The rest are left to the reader.

Proof of [1].

(1) $A \vdash B$ (by supposition).
(2) $\emptyset \vdash A \to B$.
(3) $\emptyset \vdash L(A \to B)$ (by (L+), (2)).
(4) $LA \vdash L(A \to B)$.
(5) $LA \vdash LA$.
(6) $LA \vdash LB$ (by ($\to$−(L)), (4), (5)).

Proof of [4].

(1) $L\neg A \vdash L\neg A$.
(2) $L\neg A \vdash \neg A$ (by (L−), (1)).
(3) $A \vdash \neg L\neg A$ (by (2)).
(That is, $A \vdash MA$.)

Proof of [6].

(1) $A \wedge B \vdash A, B$.
(2) $L(A \wedge B) \vdash LA, LB$ (by Thm 8.3.1 [1], (1)).
(3) $L(A \wedge B) \vdash LA \wedge LB$.

(4) $A, B \vdash A \wedge B$.
(5) $LA, LB \vdash L(A \wedge B)$ (by Thm 8.3.1 [3], (4)).
(6) $LA \wedge LB \vdash LA, LB$.
(7) $LA \wedge LB \vdash L(A \wedge B)$ (by (Tr), (6), (5)).
(8) $L(A \wedge B) \vdash\dashv LA \wedge LB$ (by (3), (7)). □

Theorem 8.3.2. (***Replaceability of equivalent formulas***)

Suppose $B \vdash\dashv_T C$ and A' results from A by replacing some (not necessarily all) occurrences of B in A by C. Then $A \vdash\dashv_T A'$.

Proof. By induction on the structure of A. The crucial step is to prove

$$A_1 \vdash\dashv_T A_1' \Longrightarrow LA_1 \vdash\dashv_T LA_1',$$

which has been established by Theorem 8.3.1 [2]. □

For simplicity we shall sometimes write "Rep eq" for the theorems of replaceability of equivalent formulas.

Theorem 8.3.3.

[1] $LA \vdash\dashv_T \neg M \neg A$.
[2] $L\neg A \vdash\dashv_T \neg MA$.
[3] $M\neg A \vdash\dashv_T \neg LA$.
[4] $LLA \vdash\dashv_T \neg MM\neg A$.
[5] $MMA \vdash\dashv_T \neg LL\neg A$.
[6] $LL\neg A \vdash\dashv_T \neg MMA$.
[7] $MM\neg A \vdash\dashv_T \neg LLA$.
[8] $LM\neg A \vdash\dashv_T \neg MLA$.
[9] $ML\neg A \vdash\dashv_T \neg LMA$.

Proof. We choose to prove [1] and [4], and the rest are left to the reader.

Proof of [1].

(1) $LA \vdash\dashv \neg\neg LA$.
(2) $\neg\neg LA \vdash\dashv \neg\neg L\neg\neg A$ (by Rep eq, $A \vdash\dashv \neg\neg A$).
(3) $LA \vdash\dashv \neg\neg L\neg\neg A$ (by (1), (2)).
(That is, $LA \vdash\dashv \neg M\neg A$.)

Proof of [4].

(1) $LLA \dashv\vdash \neg M \neg LA$ (by Thm 8.3.3 [1]).
(2) $\neg M \neg LA \dashv\vdash \neg MM \neg A$ (by Rep eq, Thm 8.3.3 [3]).
(3) $LLA \dashv\vdash \neg MM \neg A$ (by (1), (2)). □

Theorem 8.3.4.

[1] $\neg M(A \vee B) \dashv\vdash_T \neg MA \wedge \neg MB$.
[2] $M(A \vee B) \dashv\vdash_T MA \vee MB$.
[3] $L(A \to B) \vdash_T MA \to MB$.
[4] If $A \vdash_T B$, then $MA \vdash_T MB$.
[5] If $A \dashv\vdash_T B$, then $MA \dashv\vdash_T MB$.
[6] $LA \vee LB \vdash_T L(A \vee B)$.
[7] $M(A \wedge B) \vdash_T MA \wedge MB$.

Proof. We choose to prove [7]:

(1) $L\neg A \vee L\neg B \vdash L(\neg A \vee \neg B)$ (by Thm 8.3.4 [6]).
(2) $\neg L(\neg A \vee \neg B) \vdash \neg(L\neg A \vee L\neg B)$ (by (1)).
(3) $\neg L(\neg A \vee \neg B) \dashv\vdash M\neg(\neg A \vee \neg B)$ (by Thm 8.3.3 [3]).
(4) $\neg(L\neg A \vee L\neg B) \dashv\vdash \neg(\neg MA \vee \neg MB)$ (by Rep eq).
(5) $M\neg(\neg A \vee \neg B) \vdash \neg(\neg MA \vee \neg MB)$ (by (3), (2), (4)).
(6) $M\neg(\neg A \vee \neg B) \dashv\vdash M(A \wedge B)$ (by Rep eq).
(7) $\neg(\neg MA \vee \neg MB) \dashv\vdash MA \wedge MB$.
(8) $M(A \wedge B) \vdash MA \wedge MB$ (by (6), (5), (7)). □

Theorem 8.3.5.

[1] $L(\neg A \to A) \dashv\vdash_T LA$.
[2] $L(A \to \neg A) \dashv\vdash_T L\neg A$.
[3] $L(A \to B) \wedge L(\neg A \to B) \dashv\vdash_T LB$.
[4] $L(A \to B) \wedge L(A \to \neg B) \dashv\vdash_T L\neg A$.
[5] $LA \vdash_T L(B \to A)$.
[6] $L\neg A \vdash_T L(A \to B)$.
[7] $LA, MB \vdash_T M(A \wedge B)$.

Proof. We choose to prove [7]:

(1) $A, B \vdash A \wedge B$.
(2) $A \vdash B \to A \wedge B$.
(3) $LA \vdash L(B \to A \wedge B)$ (by Thm 8.3.1 [1], (2)).
(4) $L(B \to A \wedge B) \vdash MB \to M(A \wedge B)$ (by Thm 8.3.4 [3]).

(5) $LA \vdash MB \to M(A \wedge B)$ (by (3), (4)).
(6) $LA, MB \vdash M(A \wedge B)$ (by (5)). □

Theorem 8.3.6.

[1] $LA \vdash_{S_4} LLA$.
[2] $MMA \vdash_{S_4} MA$.
[3] $LA \dashv\vdash_{S_4} LLA$.
[4] $MA \dashv\vdash_{S_4} MMA$.
[5] $MLMA \vdash_{S_4} MA$.
[6] $LMA \vdash_{S_4} LMLMA$.
[7] $LMA \dashv\vdash_{S_4} LMLMA$.
[8] $MLA \dashv\vdash_{S_4} MLMLA$.

Proof. We choose to prove [8]:

(1) $LM\neg A \dashv\vdash LMLM\neg A$ (by Thm 8.3.6 [7]).
(2) $LM\neg A \dashv\vdash \neg MLA$.
(3) $LMLM\neg A \dashv\vdash \neg MLMLA$.
(4) $\neg MLA \dashv\vdash \neg MLMLA$ (by (2), (1), (3)).
(5) $MLA \dashv\vdash MLMLA$ (by (4)). □

Theorem 8.3.7.

[1] $MA \vdash_{S_5} LMA$.
[2] $MLA \vdash_{S_5} LA$.
[3] $MA \dashv\vdash_{S_5} LMA$.
[4] $LA \dashv\vdash_{S_5} MLA$.
[5] If $\Sigma \vdash_{S_5} LA$, then $\Sigma \vdash_{S_5} LLA$. (L+L)
[6] If $\Sigma \vdash_{S_5} A$, then $\Sigma \vdash_{S_5} LMA$. (LM+)

Theorem 8.3.8.

[1] $A \vdash_B LMA$.
[2] $MLA \vdash_B A$.
[3] If $MA \vdash_B B$, then $A \vdash_B LB$.

The proof of Theorems 8.3.7 and 8.3.8 is left to the reader.

Theorem 8.3.2 holds in S_4, B, and S_5 as well.

The following

1) $LLA \vdash LA,$
2) $LMA \vdash MA,$
3) $LA \vdash MLA,$
4) $MA \vdash MMA,$

hold in T, but

5) $LA \vdash LLA,$
6) $MA \vdash LMA,$
7) $MLA \vdash LA,$
8) $MMA \vdash MA,$

do not hold in T.

Since 5) and 8) hold in S_4,

9) $LA \dashv\vdash LLA,$
10) $MA \dashv\vdash MMA,$

hold in S_4, and accordingly hold in S_5.

6) and 7) hold in S_5, hence

11) $MA \dashv\vdash LMA,$
12) $LA \dashv\vdash MLA,$

hold in S_5.

9)–12) are called the *reduction laws*, which enable us to shorten certain sequences of modal symbols.

In fact, 5) is equivalent to 8) in S_4, and 6) is equivalent to 7) in S_5. Therefore S_4 can be obtained by adding the rule (L+L) to T, and S_5 can be obtained by adding (L+M) to T.

That certain rules of formal deduction do not hold in certain systems is a problem of independence (see Section 5.6).

The axiomatic deduction systems of modal systems T, S_4, S_5, and B are obtained by adding axioms and rule of inference about modal sysbols to the axiomatic deduction system of classical propositional Logic (see Section 4.1 of Chapter 4).

T contains the following two modal axioms:

$$LA \to A,$$
$$L(A \to B) \to (LA \to LB).$$

S_4, S_5, and B contain, in addition to the axioms of T, the following model axiom respectively:

$$LA \to LLA,$$
$$MA \to LMA,$$
$$A \to LMA.$$

Each of T, S_4, S_5, and B contains one rule of inference about model symbol:

From A infer LA.

The natural deduction system and axiomatic deduction system of model logic are equivalent to each other.

Exercises 8.3.

8.3.1. Prove
[1] $L(A \to B), M(A \land C) \vdash_T M(B \land C)$.
[2] $M(A \to B) \vdash\dashv_T LA \to MB$.
[3] $\emptyset \vdash_T M\neg A \lor M\neg B \lor M(A \lor B)$.

8.3.2. Prove
[1] $LA \lor LB \vdash\dashv_{S_4} L(LA \lor LB)$.
[2] $\emptyset \vdash_{S_5} L(LA \to LB) \lor L(LB \to LA)$.

8.4. SOUNDNESS

Theorem 8.4.1. (***Soundness of T***)
Suppose $\Sigma \subseteq Form(\mathcal{L}^{pm})$ and $A \in Form(\mathcal{L}^{pm})$.
[1] If $\Sigma \vdash_T A$, then $\Sigma \models_T A$.

[2] If A is T-provable, then A is T-valid.
[3] If Σ is T-satisfiable, then Σ is T-consistent.

Proof. [1] will be proved by induction on the structure of $\Sigma \vdash_T A$. Of the fourteen cases of the rules of formal deduction of the system T, only the three cases of (L−), (→−(L)), and (L+) need to be treated. The other cases are the same as in non-modal logic.

Case of (L−). We shall prove:

$$\begin{aligned} &\text{If} \quad \Sigma \models_T \text{LA}, \\ &\text{then} \quad \Sigma \models_T \text{A}. \end{aligned}$$

Suppose K is any set of T-valuations, R is any reflexive relation on K, and take any $v \in K$. Suppose $\Sigma^v = 1$. Then we have

$$(1) \qquad (\text{LA})^v = 1.$$

Since R is reflexive, we have vRv. By (1) we obtain $\text{A}^v = 1$. Hence $\Sigma \models_T \text{A}$.

Case of (→ −(L)). We shall prove:

$$\begin{aligned} &\text{If} \quad \Sigma \models_T \text{L(A} \to \text{B)}, \\ &\qquad \Sigma \models_T \text{LA}, \\ &\text{then} \quad \Sigma \models_T \text{LB}. \end{aligned}$$

Suppose K, R, and v are given as in the case of (L−), and $\Sigma^v = 1$. Then we have

$$(2) \qquad (\text{L(A} \to \text{B)})^v = (\text{LA})^v = 1.$$

Take any $v' \in K$ such that vRv'. By (2) we have $(\text{A} \to \text{B})^{v'} = \text{A}^{v'} = 1$ and then $\text{B}^{v'} = 1$. Hence $(\text{LB})^v = 1$ and $\Sigma \models_T \text{LB}$.

Case of (L+). We shall prove:

$$\begin{aligned} &\text{If} \quad \emptyset \models_T \text{A} \quad \text{(that is, A is T-valid)}, \\ &\text{then} \quad \emptyset \models_T \text{LA} \quad \text{(that is, LA is T-valid)}. \end{aligned}$$

Suppose K, R, and v are given as in the previous cases. Take any $v' \in K$ such that vRv'. Since A is T-valid, we have $\text{A}^{v'} = 1$. Hence $(\text{LA})^v = 1$ and LA is T-valid. Then [1] is proved.

[2] is a special case of [1]. [3] follows immediately from [1]. □

Theorem 8.4.2. (***Soundness of S_4***)
Suppose $\Sigma \subseteq Form(\mathcal{L}^{pm})$ and $A \in Form(\mathcal{L}^{pm})$.
[1] If $\Sigma \vdash_{S_4} A$, then $\Sigma \models_{S_4} A$.
[2] If A is S_4-provable, then A is S_4-valid.
[3] If Σ is S_4-satisfiable, then Σ is S_4-consistent.

Proof. As indicated in the proof of Theorem 8.4.1, only [1] needs to be proved. It will be proved by induction.

As in Theorem 8.4.1, the cases of rules of non-modal logic need not be treated. The cases of (L−), (→ −(L)), and (L+) can be treated as in T with modifications on the requirements of R. Hence we need to prove [1] for the rule (L+L) only, that is,

$$\text{If} \quad \Sigma \models_{S_4} \text{LA},$$
$$\text{then} \quad \Sigma \models_{S_4} \text{LLA}.$$

is to be proved. The proof is left to the reader. □

Theorem 8.4.3. (***Soundness of B***)
Suppose $\Sigma \subseteq Form(\mathcal{L}^{pm})$ and $A \in Form(\mathcal{L}^{pm})$.
[1] If $\Sigma \vdash_B A$, then $\Sigma \models_B A$.
[2] If A is B-provable, then A is B-valid.
[3] If Σ is B-satisfiable, then Σ is B-consistent. □

Theorem 8.4.4. (***Soundness of S_5***)
Suppose $\Sigma \subseteq Form(\mathcal{L}^{pm})$ and $A \in Form(\mathcal{L}^{pm})$.
[1] If $\Sigma \vdash_{S_5} A$, then $\Sigma \models_{S_5} A$.
[2] If A is S_5-provable, then A is S_5-valid.
[3] If Σ is S_5-satisfiable, then Σ is S_5-consistent.

Proof. As indicated in the proof of Theorem 8.4.2, we need to prove [1] for the rule (L+M) only, that is, we prove:

$$\text{If} \quad \Sigma \models_{S_5} \text{MA} \quad (\text{that is,} \quad \Sigma \models_{S_5} \neg\text{L}\neg\text{A}),$$
$$\text{then} \quad \Sigma \models_{S_5} \text{LMA} \quad (\text{that is,} \quad \Sigma \models_{S_5} \text{L}\neg\text{L}\neg\text{A}).$$

Suppose K is any set of S_5-valuations, R is any equivalence relation on K, and take any $v \in K$. Suppose $\Sigma^v = 1$. Then we have $(\neg\text{L}\neg\text{A})^v = 1$ and $(\text{L}\neg\text{A})^v = 0$. Hence there is some $v' \in K$ such that vRv' and

$$(1) \qquad (\neg\text{A})^{v'} = 0.$$

Take any $v'' \in K$ such that vRv''. We have $v''Rv$ because R is symmetric, and then $v''Rv'$ because R is transitive. By (1) and $v''Rv'$, we have $(\text{L}\neg\text{A})^{v''} = 0$ and accordingly

$$(2) \qquad (\neg\text{L}\neg\text{A})^{v''} = 1.$$

Then $(\text{L}\neg\text{L}\neg\text{A})^v = 1$ follows from (2) and vRv''. Hence $\Sigma \models_{\text{S}_5} \text{LMA}$.

In the above proof, S_5-valuations and values of formulas in the sense of Definition 8.2.3 are adopted. Definition 8.2.1 which is not concerned with R, may also be adopted. Then the proof proceeds as follows.

Take any set K of S_5-valuations and any $v \in K$. Suppose $\Sigma^v = 1$. We have $(\neg\text{L}\neg\text{A})^v = 1$ and $(\text{L}\neg\text{A})^v = 0$. Hence there is some $v' \in K$ such that (1) holds.

Take any $v'' \in K$. By Definition 8.2.1 and (1), we have $(\text{L}\neg\text{A})^{v''} = 0$ and $(\neg\text{L}\neg\text{A})^{v''} = 1$. Since v'' is any valuation of K, we have $(\text{L}\neg\text{L}\neg\text{A})^v = 1$ by $(\neg\text{L}\neg\text{A})^{v''} = 1$ and Definition 8.2.1. Hence $\Sigma \models_{\text{S}_5} \text{LMA}$. □

Exercises 8.4.

8.4.1. Prove the soundness of S_4.

8.4.2. Prove the soundness of B.

8.5. COMPLETENESS OF T

Lemma 8.5.1.

Suppose B, $\text{C}_1, \ldots, \text{C}_n \in \mathit{Form}(\mathcal{L}^{pm})$. If $\{\text{MB}, \text{LC}_1, \ldots, \text{LC}_n\}$ is T-consistent, then $\{\text{B}, \text{C}_1, \ldots, \text{C}_n\}$ is T-consistent.

Proof. Suppose $\{\text{MB}, \text{LC}_1, \ldots, \text{LC}_n\}$ is T-consistent, but $\{\text{B}, \text{C}_1, \ldots, \text{C}_n\}$ is not T-consistent. Then we have:

(1) $\emptyset \vdash_{\text{T}} \neg(\text{B} \wedge \text{C}_1 \wedge \ldots \wedge \text{C}_n)$.
(2) $\emptyset \vdash_{\text{T}} \text{L}\neg(\text{B} \wedge \text{C}_1 \wedge \ldots \wedge \text{C}_n)$ (by (L+), (1)).
(3) $\emptyset \vdash_{\text{T}} \neg\text{M}(\text{B} \wedge \text{C}_1 \wedge \ldots \wedge \text{C}_n)$ (by (2)).
(4) MB, $\text{L}(\text{C}_1 \wedge \ldots \wedge \text{C}_n) \vdash_{\text{T}} \text{M}(\text{B} \wedge \text{C}_1 \wedge \ldots \wedge \text{C}_n)$ (by Thm 8.3.5 [7]).
(5) $\neg\text{M}(\text{B} \wedge \text{C}_1 \wedge \ldots \wedge \text{C}_n) \vdash_{\text{T}} \neg(\text{MB} \wedge \text{L}(\text{C}_1 \wedge \ldots \wedge \text{C}_n))$ (by (4)).
(6) $\emptyset \vdash_{\text{T}} \neg(\text{MB} \wedge \text{L}(\text{C}_1 \wedge \ldots \wedge \text{C}_n))$ (by (3), (5)).
(7) $\emptyset \vdash_{\text{T}} \neg(\text{MB} \wedge \text{LC}_1 \wedge \ldots \wedge \text{LC}_n)$ (by Rep eq, Thm 8.3.1 [6]).

Since (7) contradicts the T-consistency of $\{\mathrm{MB}, \mathrm{LC}_1, \ldots, \mathrm{LC}_n\}$, we conclude that $\{\mathrm{B}, \mathrm{C}_1, \ldots, \mathrm{C}_n\}$ is T-consistent. $\square$

In the proof of the completeness of classical propositional logic we have constructed a maximal consistent set from a given consistent set of formulas (see Section 5.3). But now, to prove the completeness of the modal system T, we need to construct a system of maximal consistent sets instead of a single one. Beginning with a given T-consistent set Σ, we will construct

$$\Delta = \{\Sigma_1^*, \ldots, \Sigma_i^*, \ldots\},$$

where Σ_1^*, ..., Σ_i^*, ... are T-maximal consistent sets. We proceed as follows.

First, we extend Σ to some T-maximal consistent set Σ_1^* by adding successively all those formulas of $\mathcal{L}^{pm}$ which do not cause T-inconsistency (see the proof of Lemma 5.3.5).

Having obtained Σ_1^* we then construct for each constructed $\Sigma_i^* \in \Delta$ (including Σ_1^* itself), a series of T-maximal consistent sets. For each $\mathrm{MB} \in \Sigma_i^*$, let

$$\Sigma_j = \{\mathrm{B}\} \cup \{\mathrm{C} \mid \mathrm{LC} \in \Sigma_i^*\}.$$

We will show that Σ_j is T-consistent. Suppose $\{\mathrm{C}_1, \ldots, \mathrm{C}_n\}$ is any finite subset of Σ_j. We have $\{\mathrm{B}, \mathrm{C}_1, \ldots, \mathrm{C}_n\} \subseteq \Sigma_j$. (If B is already in $\{\mathrm{C}_1, \ldots, \mathrm{C}_n\}$, B need not be added.) Then $\{\mathrm{MB}, \mathrm{LC}_1, \ldots, \mathrm{LC}_n\} \subseteq \Sigma_i^*$. Since Σ_i^* is T-consistent, so is $\{\mathrm{MB}, \mathrm{LC}_1, \ldots, \mathrm{LC}_n\}$. By Lemma 8.5.1, $\{\mathrm{B}, \mathrm{C}_1, \ldots, \mathrm{C}_n\}$ is T-consistent. Accordingly, $\{\mathrm{C}_1, \ldots, \mathrm{C}_n\}$ is T-consistent. Thus every finite subset of Σ_j is T-consistent, and hence so is Σ_j. We extend Σ_j to some T-maximal consistent set Σ_j^* in the standard way already described. Thus for each $\mathrm{MB} \in \Sigma_i^*$ we have constructed some Σ_j^*. Each of such Σ_j^* is called a *subordinate* of Σ_i^*, written as $\Sigma_i^* sub \Sigma_j^*$.

In the foregoing paragraphs, we have described how to construct $\Delta = \{\Sigma_1^*, \ldots, \Sigma_i^*, \ldots\}$ such that for each Σ_i^* in Δ, Σ_i^* is T-maximal consistent, and for every $\mathrm{MB} \in \Sigma_i^*$, there is some T-maximal consistent set $\Sigma_j^* \in \Delta$ such that $\Sigma_i^* sub \Sigma j^*$, $\mathrm{B} \in \Sigma_j^*$, and $\mathrm{C} \in \Sigma_j^*$ for every $\mathrm{LC} \in \Sigma_i^*$.

Now, for every $\Sigma_i^* \in \Delta$, we construct a valuation v_i such that $\mathrm{p}^{v_i} = 1$ iff $\mathrm{p} \in \Sigma_i^*$ for every proposition symbol p. Let

$$K = \{v_i \mid \Sigma_i^* \in \Delta\},$$

and let R be a binary relation on K such that $v_i R v_j$ iff $\Sigma_i^* = \Sigma_j^*$ or $\Sigma_i^* sub \Sigma_j^*$ (for every v_i, $v_j \in K$). Then R is reflexive.

These conventions stated above will be used throughout this section.

Lemma 8.5.2.
Suppose Σ_i^*, $\Sigma_j^* \in \Delta$ such that $\Sigma_i^* = \Sigma_j^*$ or $\Sigma_i^* sub \Sigma_j^*$, and suppose $\mathrm{LB} \in \Sigma_i^*$. Then $\mathrm{B} \in \Sigma_j^*$.

Proof. We distinguish between two cases. For the first case, $\Sigma_i^* = \Sigma_j^*$, we have the following:

$$\mathrm{LB} \vdash_{\mathrm{T}} \mathrm{B}.$$
$$\emptyset \vdash_{\mathrm{T}} \mathrm{LB} \to \mathrm{B}.$$
$$\Sigma_i^* \vdash_{\mathrm{T}} \mathrm{LB} \to \mathrm{B}.$$
$$\mathrm{LB} \to \mathrm{B} \in \Sigma_i^* \quad \text{(by T-max consis of } \Sigma_i^*\text{)}.$$
$$\mathrm{LB} \in \Sigma_i^* \quad \text{(by supposition)}.$$
$$\mathrm{B} \in \Sigma_i^* \quad \text{(by Lem 5.3.3)}.$$
$$\mathrm{B} \in \Sigma_j^* \quad \text{(by } \Sigma_i^* = \Sigma_j^*\text{)}.$$

For the second case, $\Sigma_i^* sub \Sigma_j^*$, by $\mathrm{LB} \in \Sigma_i^*$ and the construction of Σ_j^*, we have $\mathrm{B} \in \Sigma_j^*$. □

Lemma 8.5.3.
For every $\mathrm{A} \in Form(\mathcal{L}^{pm})$ and every $v_i \in K$, $\mathrm{A}^{v_i} = 1$ iff $\mathrm{A} \in \Sigma_i^*$.

Proof. By induction on the structure of A. The cases of A being an atom, $\neg\mathrm{B}$, $\mathrm{B} \wedge \mathrm{C}$, $\mathrm{B} \vee \mathrm{C}$, $\mathrm{B} \to \mathrm{C}$, or $\mathrm{B} \leftrightarrow \mathrm{C}$ are routine and are left to the reader. We need to prove for the case of $\mathrm{A} = \mathrm{LB}$.

First, we shall prove $\mathrm{LB} \in \Sigma_i^* \implies (\mathrm{LB})^{v_i} = 1$. Suppose $\mathrm{LB} \in \Sigma_i^*$. Take any $v_j \in K$ such that $v_i R v_j$. Then $\Sigma_i^* = \Sigma_j^*$ or $\Sigma_i^* sub \Sigma_j^*$. We have $\mathrm{B} \in \Sigma_j^*$ (by Lemma 8.5.2) and $\mathrm{B}^{v_j} = 1$ (by the induction hypothesis). Hence $(\mathrm{LB})^{v_i} = 1$.

Then for the converse, suppose $(\mathrm{LB})^{v_i} = 1$ and $\mathrm{LB} \notin \Sigma_i^*$. We have

$$\neg\mathrm{LB} \in \Sigma_i^* \quad \text{(by T-max consis of } \Sigma_i^*\text{)}.$$
$$\neg\mathrm{LB} \vdash_{\mathrm{T}} \mathrm{M}\neg\mathrm{B}$$
$$\emptyset \vdash_{\mathrm{T}} \neg\mathrm{LB} \to \mathrm{M}\neg\mathrm{B}$$
$$\Sigma_i^* \vdash_{\mathrm{T}} \neg\mathrm{LB} \to \mathrm{M}\neg\mathrm{B}$$
$$\neg\mathrm{LB} \to \mathrm{M}\neg\mathrm{B} \in \Sigma_i^* \quad \text{(by T-max consis of } \Sigma_i^*\text{)}.$$
$$\mathrm{M}\neg\mathrm{B} \in \Sigma_i^* \quad \text{(by Lem 5.3.3)}.$$

By $M\neg B \in \Sigma_i^*$ there is some $\Sigma_j^* \in \Delta$ such that $\Sigma_i^* sub \Sigma_j^*$ (and hence $v_i R v_j$) and $\neg B \in \Sigma_j^*$. Then we obtain

$$B \notin \Sigma_j^* \quad \text{(by T-max consis of } \Sigma_j^*\text{).}$$
$$B^{v_j} = 0 \quad \text{(by ind hyp).}$$

That is, there is some $v_j \in K$ such that $v_i R v_j$ and $B^{v_j} = 0$. Hence $(LB)^{v_i} = 0$, yielding a contradiction, and we have $LB \in \Sigma_i^*$. □

Lemma 8.5.3 is crucial in the proof of completeness. It is analogous to Lemma 5.3.6 in the proof of the completeness of non-modal propositional logic.

Theorem 8.5.4. (***Completeness of T***)
Suppose $\Sigma \subseteq Form(\mathcal{L}^{pm})$ and $A \in Form(\mathcal{L}^{pm})$.

[1] If Σ is T-consistent, then Σ is T-satisfiable.
[2] If $\Sigma \models_T A$, then $\Sigma \vdash_T A$.
[3] If A is T-valid, then A is T-provable.

Proof. Suppose Σ is T-consistent. Σ can be extended to some T-maximal consistent set $\Sigma_1^* \in \Delta$ as described above. Take any $A \in \Sigma$. We have $A \in \Sigma_1^*$. By Lemma. 8.5.3, $A^{v_1} = 1$. Hence Σ is T-satisfiable and [1] is proved.

[2] follows immediately from [1], and [3] is a special case of [2]. □

8.6. COMPLETENESS OF S_4, B, S_5

The proof of the completeness of S_4, B, and S_5 is essentially analogous to that of T, because the distinction between the semantics of these systems and that of T consists only in the different requirements of the relation R on the set K of valuations.

Given any S_4 (B, S_5)-consistent set Σ of formulas, we first extend Σ (in the same way as in the case of T) to some S_4 (B, S_5)-maximal consistent set Σ_1^*, and then construct a series of Σ_j^* from every constructed $S_4(B, S_5)$-maximal consistent Σ_i^* such that Σ_j^* is S_4 (B, S_5)-maximal consistent and $\Sigma_i^* sub \Sigma_j^*$. Then we obtain $\Delta = \{\Sigma_1^*, \ldots, \Sigma_i^*, \ldots\}$.

For $n \geq 1$, we define $\Sigma_i^* sub_n \Sigma_j^*$ as follows:

$$\Sigma_i^* sub_1 \Sigma_j^* \quad \text{iff} \quad \Sigma_i^* sub \Sigma_j^*.$$

$$\Sigma_i^* sub_{k+1} \Sigma_j^* \quad \text{iff} \quad \text{there is some} \quad \Sigma_r^* \in \Delta \quad \text{such that}$$
$$\Sigma_i^* sub_k \Sigma_r^* \quad \text{and} \quad \Sigma_r^* sub \Sigma_j^*.$$

For every $\Sigma_i^* \in \Delta$, we construct a valuation v_i such that for every proposition symbol p, $\mathrm{p}^{v_i} = 1$ iff $\mathrm{p} \in \Sigma_i^*$. Let $K = \{v_i \mid \Sigma_i^* \in \Delta\}$. These conventions are the same as those in the last section.

In the following, we will first formulate and prove the lemmas for S_4, B, and S_5 which correspond to Lemmas 8.5.2 and 8.5.3 for T, and then state simultaneously the completeness theorems for S_4, B, and S_5.

The following Lemmas 8.6.1 and 8.6.2 are for S_4, where R is supposed to be a binary relation on K such that for any $v_i, v_j \in K$, $v_i R v_j$ iff $\Sigma_i^* = \Sigma_j^*$ or $\Sigma_i^* sub_n \Sigma_j^*$ for some $n \geq 1$. Then, R is reflexive and transitive.

Lemma 8.6.1.

Suppose Σ_i^*, $\Sigma_j^* \in \Delta$ such that $\Sigma_i^* = \Sigma_j^*$ or $\Sigma_i sub_n \Sigma_j^*$ for some $n \geq 1$, and suppose LB $\in \Sigma_i^*$. Then B $\in \Sigma_j^*$.

Proof. We distinguish between two cases. In the first case, $\Sigma_i^* = \Sigma_j^*$, we have B $\in \Sigma_j^*$ by Lemma 8.5.2.

In the second case, $\Sigma_i^* sub_n \Sigma_j^*$ for some $n \geq 1$. Then B $\in \Sigma_j^*$ is to be proved by induction on n.

Basis. $\Sigma_i^* sub_1 \Sigma_j^*$ means $\Sigma_i^* sub \Sigma_j^*$, then B $\in \Sigma_j^*$ is obtained by Lemma 8.5.2.

Induction step. $\Sigma_i^* sub_{k+1} \Sigma_j^*$ means that there is some $\Sigma_r^* \in \Delta$ such that $\Sigma_i^* sub_k \Sigma_r^*$ and $\Sigma_r^* sub \Sigma_j^*$. Then we have the following:

$$\mathrm{LB} \vdash_{S_4} \mathrm{LLB}.$$
$$\emptyset \vdash_{S_4} \mathrm{LB} \rightarrow \mathrm{LLB}.$$
$$\Sigma_i^* \vdash_{S_4} \mathrm{LB} \rightarrow \mathrm{LLB}.$$
$$\mathrm{LB} \rightarrow \mathrm{LLB} \in \Sigma_i^* \quad (\text{by } S_4\text{-max consis of } \Sigma_i^*).$$
$$\mathrm{LB} \in \Sigma_i^* \quad (\text{by supposition}).$$
$$\mathrm{LLB} \in \Sigma_i^* \quad (\text{by Lem 5.3.3}).$$
$$\mathrm{LB} \in \Sigma_r^* \quad (\text{by } \Sigma_i^* sub_k \Sigma_r^*, \text{ ind hyp}).$$

By LB $\in \Sigma_r^*$, $\Sigma_r^* sub \Sigma_j^*$, and Lemma 8.5.2, we derive B $\in \Sigma_j^*$. □

Lemma 8.6.2.

For every A $\in Form(\mathcal{L}^{pm})$ and every $v_i \in K$, $\mathrm{A}^{v_i} = 1$ iff A $\in \Sigma_i^*$.

Proof. By induction on the structure of A. As indicated in the proof of Lemma 8.5.3, we need to prove for the case of A = LB only.

First, we shall prove $\mathrm{LB} \in \Sigma_i^* \Longrightarrow (\mathrm{LB})^{v_i} = 1$. Suppose $\mathrm{LB} \in \Sigma_i^*$. Take any $v_j \in K$ such that $v_i R v_j$. We have $\Sigma_i^* = \Sigma_j^*$ or $\Sigma_i^* sub_n \Sigma_j^*$ for some $n \geq 1$. Then we obtain $\mathrm{B} \in \Sigma_j^*$ (by Lemma 8.6.1) and $\mathrm{B}^{v_j} = 1$ (by the induction hypothesis). Hence $(\mathrm{LB})^{v_i} = 1$.

The converse is to be proved in the same way as in the case of T in Lemma 8.5.3. □

The following Lemmas 8.6.3 and 8.6.4 are for B, where R is suppose to be a binary relation on K such that $v_i R v_j$ iff $\Sigma_i^* = \Sigma_j^*$ or $\Sigma_i^* sub \Sigma_j^*$ or $\Sigma_j^* sub \Sigma_i^*$. Then R is reflexive and symmetric.

Lemma 8.6.3.
Suppose $\Sigma_i^*, \Sigma_j^* \in \Delta$ such that $\Sigma_i^* = \Sigma_j^*$ or $\Sigma_i^* sub \Sigma_j^*$ or $\Sigma_j^* sub \Sigma_i^*$, and suppose $\mathrm{LB} \in \Sigma_i^*$. Then $\mathrm{B} \in \Sigma_j^*$.

Proof. In the cases of $\Sigma_i^* = \Sigma_j^*$ or $\Sigma_i^* sub \Sigma_j^*$, we obtain $\mathrm{B} \in \Sigma_j^*$ by Lemma 8.5.2.

In the case of $\Sigma_j sub \Sigma_i^*$, suppose $\mathrm{B} \notin \Sigma_j^*$. Then we have

$$\neg \mathrm{B} \in \Sigma_j^* \quad \text{(by B-max consis of } \Sigma_j^*\text{).}$$
$$\neg \mathrm{B} \vdash_{\mathrm{B}} \mathrm{LM}\neg \mathrm{B} \quad \text{(by (LM+)).}$$
$$\neg \mathrm{B} \vdash_{\mathrm{B}} \mathrm{L}\neg\mathrm{LB} \quad \text{(by Rep eq, } \mathrm{M}\neg\mathrm{B} \vdash\dashv_{\mathrm{B}} \neg\mathrm{LB}\text{).}$$
$$\emptyset \vdash_{\mathrm{B}} \neg\mathrm{B} \to \mathrm{L}\neg\mathrm{LB}.$$
$$\Sigma_j^* \vdash_{\mathrm{B}} \neg\mathrm{B} \to \mathrm{L}\neg\mathrm{LB}.$$
$$\neg\mathrm{B} \to \mathrm{L}\neg\mathrm{LB} \in \Sigma_j^* \quad \text{(by B-max consis of } \Sigma_j^*\text{).}$$
$$\mathrm{L}\neg\mathrm{LB} \in \Sigma_j^* \quad \text{(by Lem 5.3.3).}$$
$$\neg\mathrm{LB} \in \Sigma_i^* \quad \text{(by } \Sigma_j^* sub \Sigma_i^*\text{).}$$

By $\neg\mathrm{LB} \in \Sigma_i^*$ and the supposition $\mathrm{LB} \in \Sigma_i^*$, Σ_i^* is B-inconsistent, contradicting the B-maximal consistency of Σ_i^*. Hence $\mathrm{B} \in \Sigma_j^*$. □

Lemma 8.6.4.
For every $\mathrm{A} \in Form(\mathcal{L}^{pm})$ and every $v_i \in K$, $\mathrm{A}^{v_i} = 1$ iff $\mathrm{A} \in \Sigma_i^*$.

Proof. By induction on the structure of A. As stated before, we need to prove for the case of A = LB only.

First, we shall prove $LB \in \Sigma_i^* \Longrightarrow (LB)^{v_i} = 1$. Suppose $LB \in \Sigma_i^*$. Take any $v_j \in K$ such that $v_i R v_j$. We have $\Sigma_i^* = \Sigma_j^*$ or $\Sigma_i^* sub \Sigma_j^*$ or $\Sigma_j^* sub \Sigma_i^*$. We then obtain $B \in \Sigma_j^*$ (by Lemma 8.6.3) and $B^{v_j} = 1$ (by the induction hypothesis). Hence $(LB)^{v_i} = 1$.

The converse will be proved in the same way as in the case of T in Lemma 8.5.3. □

The following Lemmas 8.6.5 and 8.6.6 are for S_5, where R is suppose to be a binary relation on K such that $v_i R v_j$ iff $\Sigma_i^* = \Sigma_j^*$ or $\Sigma_i^* sub_n \Sigma_j^*$ for some $n \geq 1$ or $\Sigma_j^* sub \Sigma_i^*$. Then R is an equivalence relation.

Lemma 8.6.5.

Suppose $\Sigma_i^*, \Sigma_j^* \in \Delta$ such that $\Sigma_i^* = \Sigma_j^*$ or $\Sigma_i^* sub_n \Sigma_j^*$ for some $n \geq 1$ or $\Sigma_j^* sub \Sigma_i^*$, and suppose $LB \in \Sigma_i^*$. Then $B \in \Sigma_j^*$.

Proof. In the cases of $\Sigma_i^* = \Sigma_j^*$ or $\Sigma_i^* sub_n \Sigma_j^*$ for some $n \geq 1$, $B \in \Sigma_j^*$ is obtained by Lemma 8.6.1.

In the case of $\Sigma_j^* sub \Sigma_i^*$, $B \in \Sigma_j^*$ will be obtained as in Lemma 8.6.3, because the rule (LM+) of the system B, which is used in proving Lemma 8.6.3, holds in S_5. □

Lemma 8.6.6.

For every $A \in Form(\mathcal{L}^{pm})$ and every $v_i \in K$, $A^{v_i} = 1$ iff $A \in \Sigma_i^*$.

Proof. By induction on the structure of A. We need to prove for the case of $A = LB$ only.

First, we shall prove $LB \in \Sigma_i^* \Longrightarrow (LB)^{v_i} = 1$. Suppose $LB \in \Sigma_i^*$. Take any $v_j \in K$ such that $v_i R v_j$. We have $\Sigma_i^* = \Sigma_j^*$ or $\Sigma_i^* sub_n \Sigma_j^*$ for some $n \geq 1$ or $\Sigma_j^* sub \Sigma_i^*$. Then we obtain $B \in \Sigma_j^*$ (by Lemma 8.6.5) and $B^{v_j} = 1$ (by the induction hypothesis). Hence $(LB)^{v_i} = 1$.

The converse can be proved in the same way as in Lemma 8.5.3. □

Lemmas 8.6.5 and 8.6.6 for S_5 are based on the semantics formulated in Definition 8.2.3, which is concerned with an equivalence relation. Lemma 8.6.6 can be re-established (see Lemma 8.6.8) according to the semantics formulated in Definition 8.2.1 which is not concerned with any equivalence relation.

The following Lemmas 8.6.7 and 8.6.8 are for S_5.

Lemma 8.6.7.
Suppose Σ_i^*, $\Sigma_j^* \in \Delta$ such that $\Sigma_i^* = \Sigma_j^*$ or $\Sigma_i^* sub_n \Sigma_j^*$ for some $n \geq 1$, and suppose LB $\in \Sigma_j^*$. Then B $\in \Sigma_i^*$.

Proof. The first case is $\Sigma_i^* = \Sigma_j^*$. By Lemma 8.5.2, B $\in \Sigma_i^*$ is derived from LB $\in \Sigma_j^*$.

In the second case, $\Sigma_i^* sub_n \Sigma_j^*$ for some $n \geq 1$, B $\in \Sigma_i^*$ is to be proved by induction on n.

Basis. $\Sigma_i^* sub_1 \Sigma_j^*$ means $\Sigma_i^* sub \Sigma_j^*$. Then B $\in \Sigma_i^*$ follows from LB $\in \Sigma_j^*$ as in the third case in the proof of Lemma 8.6.3.

Induction step. $\Sigma_i^* sub_{k+1} \Sigma_j^*$ means that there is some $\Sigma_r^* \in \Delta$ such that $\Sigma_i^* sub_k \Sigma_r^*$ and $\Sigma_r^* sub \Sigma_j^*$. Then we have

$$\text{LB} \in \Sigma_j^* \quad \text{(by supposition).}$$
$$\emptyset \vdash_{S_5} \text{LB} \to \text{LLB} \quad \text{(by (L+L)).}$$
$$\Sigma_j^* \vdash_{S_5} \text{LB} \to \text{LLB.}$$
$$\text{LB} \to \text{LLB} \in \Sigma_j^* \quad \text{(by S}_5\text{-max consis of } \Sigma_j^*\text{).}$$
$$\text{LLB} \in \Sigma_j^* \quad \text{(by Lem 5.3.3).}$$
$$\text{LB} \in \Sigma_r^* \quad \text{(by } \Sigma_r^* sub \Sigma_j^*\text{, basis).}$$

By LB $\in \Sigma_r^*$, $\Sigma_i^* sub_k \Sigma_r^*$, and the induction hypothesis, we obtain B $\in \Sigma_i^*$.
□

Lemma 8.6.8.
For every A $\in Form(\mathcal{L}^{pm})$ and every $v_i \in K$, $\text{A}^{v_i} = 1$ iff A $\in \Sigma_i^*$.

Proof. By induction on the structure of A. We need to prove for the case of A = LB only.

Now we will first prove $(\text{LB})^{v_i} = 1 \Longrightarrow \text{LB} \in \Sigma_i^*$. Suppose $(\text{LB})^{v_i} = 1$ but LB $\notin \Sigma_i^*$. Then we have the following:

$$\neg\text{LB} \in \Sigma_i^* \quad \text{(by S}_5\text{-max consis of } \Sigma_i^*\text{).}$$
$$\neg\text{LB} \vdash_{S_5} \text{M}\neg\text{B}$$
$$\emptyset \vdash_{S_5} \neg\text{LB} \to \text{M}\neg\text{B}$$
$$\Sigma_i^* \vdash_{S_5} \neg\text{LB} \to \text{M}\neg\text{B.}$$
$$\neg\text{LB} \to \text{M}\neg\text{B} \in \Sigma_i^* \quad \text{(by S}_5\text{-max consis of } \Sigma_i^*\text{).}$$
$$\text{M}\neg\text{B} \in \Sigma_i^* \quad \text{(by Lem 5.3.3).}$$

By $M\neg B \in \Sigma_i^*$, there is some $\Sigma_j^* \in \Delta$ such that $\Sigma_i^* sub \Sigma_j^*$ and $\neg B \in \Sigma_j^*$. We have $B \notin \Sigma_j^*$ (by the S_5-maximal consistency of Σ_j^*) and $B^{v_j} = 0$ (by the induction hypothesis). Then $(LB)^{v_i} = 0$, contradicting $(LB)^{v_i} = 1$. Hence $LB \in \Sigma_i^*$.

For the converse, suppose $LB \in \Sigma_i^*$. We have

$$LB \to LLB \in \Sigma_i^* \quad \text{(by } S_5\text{-max consis of } \Sigma_i^*\text{)}.$$
$$LLB \in \Sigma_i^* \quad \text{(by Lem 5.3.3)}.$$

Since $\Sigma_1^* sub_n \Sigma_i^*$ for some $n \geq 1$, we obtain $LB \in \Sigma_1^*$ by Lemma 8.6.7.

Take any $v_j \in K$. We have $\Sigma_1^* sub_m \Sigma_j^*$ for some $m \geq 1$. By Lemma 8.6.1, we obtain $B \in \Sigma_j^*$ from $LB \in \Sigma_1^*$. Then we derive $B^{v_j} = 1$ by the induction hypothesis. Hence $(LB)^{v_i} = 1$. □

Theorem 8.6.9. (***Completeness of*** S_4, B, S_5)
Suppose $\Sigma \subseteq Form(\mathcal{L}^{pm})$, $A \in Form(\mathcal{L}^{pm})$.
[1] If Σ is S_4 (B, S_5)-consistent, then Σ is S_4 (B, S_5)-satisfiable.
[2] If $\Sigma \models_{S_4(B,S_5)} A$, then $\Sigma \vdash_{S_4(B,S_5)} A$.
[3] If A is S_4 (B, S_5)-valid, then A is S_4 (B, S_5)-provable. □

9

MODAL FIRST-ORDER LOGIC

Modal first-order logic is constructed by adding modal notions to classical first-order logic, in essentially the same way as modal propositional logic is constructed from classical propositional logic. But the situations with modal first-order logic are more complicated. We shall construct various systems of modal first-order logic corresponding to the systems T, S_4, S_5, and B.

9.1. MODAL FIRST-ORDER LANGUAGE

The *modal first-order language* $\mathcal{L}^m$ is obtained by adding the necessity symbol L to the first-order language $\mathcal{L}$.

The sets $Term(\mathcal{L}^m)$ and $Atom(\mathcal{L}^m)$ of terms and atoms of $\mathcal{L}^m$ are the same as $Term(\mathcal{L})$ and $Atom(\mathcal{L})$ respectively.

The set $Form(\mathcal{L}^m)$ of formulas of $\mathcal{L}^m$ is defined to be the smallest class of expressions of $\mathcal{L}^m$ closed under the following formation rules of formulas of $\mathcal{L}^m$:

[1] $Atom(\mathcal{L}^m) \subseteq Form(\mathcal{L}^m)$.

[2] If $A \in Form(\mathcal{L}^m)$, then $(\neg A)$, $(LA) \in Form(\mathcal{L}^m)$.

[3] If A, B $\in Form(\mathcal{L}^m)$, then $(A * B) \in Form(\mathcal{L}^m)$, $*$ being any one of $\wedge$, $\vee$, $\rightarrow$, $\leftrightarrow$.

[4] If $A(u) \in Form(\mathcal{L}^m)$, x not occurring in A(u), then $\forall xA(x)$, $\exists xA(x)$ $\in Form(\mathcal{L}^m)$.

The details of the structure of formulas of $\mathcal{L}^m$ are left to the reader.

The systems of modal first-order logic corresponding to the systems T, S_4, S_5, and B are TQ, S_4Q, S_5Q, and BQ respectively, where Q means quantificational or with quantification. Hence TQ, for instance, is the system T with quantification.

Other systems of modal first-order logic corresponding to T, S_4, S_5, and B are TQ(BF), S_4Q(BF), S_5Q(BF), and BQ(BF). The meaning of BF will be explained in the next section.

In the following sections, modal first-order logic without equality will be considered first, and systems with equality will be studied later in Section 9.6.

9.2. SEMANTICS

Essentially, the semantics for modal first-order logic is constructed by combining those for modal propositional logic and classical first-order logic. But we are faced with the following question. In the case of classical logic, the value of formulas under a certain valuation v is concerned only with v itself, while in modal logic it is concerned with a set K of valuations or with certain valuations in K, of which v is a member. Then we may have different domains associated with different valuations in K or have a single domain for all valuations in K. We will consider the first case in Definition 9.2.1.

Definition 9.2.1. (***Valuation***)

Suppose K is a set and R is a reflexive relation on K. Each element $v \in K$ is called a *TQ-valuation* for $\mathcal{L}^m$, which consists of a domain $D(v)$ assigned peculiarly to v and a function (denoted by v) with the set of all non-logical symbols and free variable symbols as domain such that

[1] If $v, v' \in K$ and vRv', then $D(v) \subseteq D(v')$.

[2] $\mathrm{a}^v, \mathrm{u}^v \in D(v)$, a and u being any individual symbol and free variable symbol respectively.

[3] $\mathrm{F}^v \subseteq D(v)^n$, F being any n-ary relation symbol.

[4] $\mathrm{f}^v : D(v)^n \to D(v)$, f being any n-ary function symbol.

$S_4Q(S_5Q, BQ)$-valuations are defined analogously by making the familiar modifications of the requirements of R such that R be reflexive and transitive for S_4Q, be an equivalence relation for S_5Q, and be reflexive and

symmetric for BQ. (S_5Q-*valuation* can also be defined independently of R. This is left to the reader.)

Definition 9.2.2. (***Value of terms and formulas***)

Suppose K and R are given as in Definition 9.2.1. The *value* of terms under valuation $v \in K$ is defined by recursion:

[1] $\mathrm{a}^v,\ \mathrm{u}^v \in D(v)$.

[2] $\mathrm{f}(\mathrm{t}_1, \ldots, \mathrm{t}_n)^v = \mathrm{f}^v(\mathrm{t}_1^v, \ldots, \mathrm{t}_n^v)$.

The *value* of formulas under valuation $v \in K$ is defined by recursion:

[1] $\mathrm{F}(\mathrm{t}_1, \ldots, \mathrm{t}_n)^v = \begin{cases} 1 & \text{if } \langle \mathrm{t}_1^v, \ldots, \mathrm{t}_n^v \rangle \in \mathrm{F}^v, \\ 0 & \text{otherwise.} \end{cases}$

[2]–[7] Same as in Definition 8.2.3.

[8] $\forall \mathrm{x}\mathrm{A}(\mathrm{x})^v = \begin{cases} 1 & \text{if for every } \alpha \in D(v),\ \mathrm{A}(\mathrm{u})^{v(\mathrm{u}/\alpha)} = 1, \\ & \quad \text{u not occurring in } \mathrm{A}(\mathrm{x}), \\ 0 & \text{otherwise.} \end{cases}$

[9] $\exists \mathrm{x}\mathrm{A}(\mathrm{x})^v = \begin{cases} 1 & \text{if for some } \alpha \in D(v),\ \mathrm{A}(\mathrm{u})^{v(\mathrm{u}/\alpha)} = 1, \\ & \quad \text{u not occurring in } \mathrm{A}(\mathrm{x}), \\ 0 & \text{otherwise.} \end{cases}$

Definition 9.2.3. (***Satisfiability, validity***)

$\Sigma \subseteq Form(\mathcal{L}^m)$ is *TQ-satisfiable* iff for some set K of TQ-valuations, some reflexive relation R on K, and some $v \in K$, $\Sigma^v = 1$.

$\mathrm{A} \in Form(\mathcal{L}^m)$ is *TQ-valid* iff for every set K of TQ-valuations, every reflexive relation R on K, and every $v \in K$, $\mathrm{A}^v = 1$.

S_4Q (S_5Q, BQ)-*satisfiability* and S_4Q (S_5Q, BQ)-*validity* are defined analogously with modifications of the requirements of R. (S_5Q-*satisfiability* and S_5Q-*validity* can be defined independently of R.)

Logical consequences $\Sigma \models_{\mathrm{TQ}} \mathrm{A}$, $\Sigma \models_{\mathrm{S_4Q}} \mathrm{A}$, $\Sigma \models_{\mathrm{S_5Q}} \mathrm{A}$, and $\Sigma \models_{\mathrm{BQ}} \mathrm{A}$ are defined as in non-modal systems with suitable modifications.

Now we turn to the second case of the semantics of modal first-order logic mentioned at the begining of this section, in which we have one single domain for all the valuations in K. Replacing $D(v)$ for each $v \in K$ by a single domain D and deleting the requirements of $D(v)$ in Definition 9.2.1, we obtain new definitions of valuations, satisfiability, and validity, formulated as follows.

Definition 9.2.4. (***Valuation***)

Suppose K is a set and R is a reflexive relation on K. Each element $v \in K$ is called a *TQ(BF)-valuation* for $\mathcal{L}^m$, which consists of a domain D (which is available for every valuation in K) and a function (denoted by v) with the set of all non-logical symbols and free variable symbols as domain such that

[1] $\mathrm{a}^v, \mathrm{u}^v \in D$.

[2] $\mathrm{F}^v \subseteq D^n$.

[3] $\mathrm{f}^v : D^n \to D$.

$S_4Q(BF)$ ($S_5Q(BF)$, $BQ(BF)$)-*valuations* are defined analogously with modifications of the requirements of R. ($S_5Q(BF)$-*valuation* can also be defined independently of R.)

The *value* of terms and formulas under valuation $v \in K$ will be defined in the same way as in Definition 9.2.2 except that D is used instead of $D(v)$.

Then, $TQ(BF)$ ($S_4Q(BF)$, $S_5Q(BF)$, $BQ(BF)$)-*satisfiability* and $TQ(BF)$ ($S_4Q(BF)$, $S_5Q(BF)$, $BQ(BF)$)-*validity* are defined in terms of $TQ(BF)$ ($S_4Q(BF)$, $S_5Q(BF)$, $BQ(BF)$)-valuations in the same way as in Definition 9.2.3.

$\Sigma \models_{\mathrm{TQ(BF)}} \mathrm{A}$, $\Sigma \models_{\mathrm{S_4Q(BF)}} \mathrm{A}$, $\Sigma \models_{\mathrm{S_5Q(BF)}} \mathrm{A}$, and $\Sigma \models_{\mathrm{BQ(BF)}} \mathrm{A}$ are defined in a similar way.

The distinction between these two kinds of valuations defined in Definitions 9.2.1 and 9.2.4 can be explained by means of the formula

$$\text{BF} \qquad \forall \mathrm{x}\mathrm{L}\mathrm{A}(\mathrm{x}) \to \mathrm{L}\forall \mathrm{x}\mathrm{A}(\mathrm{x})$$

which is named the *Barcan formula*, due to Ruth C. Barcan.

According to the semantics based on Definition 9.2.4, BF is TQ(BF) ($\mathrm{S_4Q(BF)}, \mathrm{S_5Q(BF)}, \mathrm{BQ(BF)}$)-valid. But according to the semantics based on Definition 9.2.1, BF is not TQ-valid nor is it $\mathrm{S_4Q}$-valid. These assertions are demonstrated as follows.

Suppose K is any set of TQ(BF)-valuations, R is any reflexive relation on K. Take any $v \in K$ over domain D. Suppose $\forall \mathrm{xLA(x)}^v = 1$. Then, for every $\alpha \in D$ and every $v' \in K$ such that vRv', we have $v(\mathrm{u}/\alpha)Rv'(\mathrm{u}/\alpha)$ and

$$(\mathrm{LA(u)})^{v(\mathrm{u}/\alpha)} = 1, \text{ u not occurring in A(x);}$$

$$\mathrm{A(u)}^{v'(\mathrm{u}/\alpha)} = 1;$$

$$\forall \mathrm{xA(x)}^{v'} = 1;$$

$$(\mathrm{L}\forall \mathrm{xA(x)})^v = 1.$$

Hence $\forall xLA(x) \to L\forall xA(x)$ is valid in TQ(BF). Similarly for S_4Q(BF), S_5Q(BF), or BQ(BF).

To refute the TQ(S_4Q)-validity of BF, we take an instance of BF:

$$\forall xLF(x) \to L\forall xF(x),$$

where F is a unary relation symbol. Suppose $K = \{v_1, v_2\}$ and R is a binary relation on K such that

1) $$v_1Rv_1,\ v_1Rv_2,\ v_2Rv_2,\ \text{not } v_2Rv_1.$$

Then R is reflexive and transitive, but not symmetric. Suppose

$$D(v_1) = \{\alpha\},$$
$$D(v_2) = \{\alpha, \beta\}.$$

We have $D(v_1) \subseteq D(v_2)$. Take any free variable symbol u and let

$$u^{v_1} = \alpha,$$
$$u^{v_2} = \beta,$$
$$F^{v_1} = \{\alpha\},$$
$$F^{v_2} = \{\beta\}.$$

Then we have

2) $$F(u)^{v_1} = F(u)^{v_2} = 1.$$

3) $$\forall xF(x)^{v_2} = 0.$$

By 2) and 1), we obtain $(LF(u))^{v_1} = 1$. Since $D(v_1)$ contains only one member α, we have $\forall xLF(x)^{v_1} = 1$. Since v_1Rv_2, we derive $(L\forall xF(x))^{v_1} = 0$ by 3). Hence $(\forall xLF(x) \to L\forall xF(x))^{v_1} = 0$. Since R is reflexive and transitive, BF is not TQ-valid nor is it S_4Q-valid.

Note that since R is not symmetric, we have not refuted the S_5Q(BQ)-validity by the above arguments. In fact, BF is S_5Q-valid and BQ-valid.

9.3. FORMAL DEDUCTION

The rules of formal deduction of TQ include the three rules (L$-$), ($\to$ $-$(L)), and (L+) (which are added to classical propositional logic to obtain

T) in addition to those of classical first-order logic, with the equality symbol not considered for the time being.

Then the rules of S_4Q, S_5Q, and BQ are obtained by adding, respectively, the rules $(L+L)$, $(L+M)$, and (LM+) to those of TQ.

By the soundness of TQ and S_4Q (see Theorem 9.4.1 in the next section, which is based on the semantics formulated in Definition 9.2.1) and the TQ-invalidity and S_4Q-invalidity of BF (see the proof in the last section), BF is not formally provable in TQ or S_4Q.

But BF is formally provable in S_5Q and BQ. The formal proof is as follows:

1) $\forall xLA(x) \vdash LA(u)$ (take u not occurring in A(x)).
2) $M\forall xLA(x) \vdash MLA(u)$ (by Thm 8.3.4 [4], 1)).
3) $MLA(u) \vdash A(u)$ (by Thm 8.3.8 [2]).
4) $M\forall xLA(x) \vdash A(u)$ (by 2), 3)).
5) $M\forall xLA(x) \vdash \forall xA(x)$ (by 4)).
6) $\forall xLA(x) \vdash L\forall xA(x)$ (by Thm 8.3.8 [3], 5)).
7) $\emptyset \vdash \forall xLA(x) \to L\forall xA(x)$ (by 6)).

The formula BF can be formulated as a rule of formal deduction:

(BF) If $\Sigma \vdash \forall xLA(x)$,
then $\Sigma \vdash L\forall xA(x)$.

Then by the foregoing explanations we may add the rule (BF) to TQ and S_4Q to obtain stronger systems TQ(BF) and S_4Q(BF). But (BF) need not be added to S_5Q or BQ, because it can be derived in these systems. That is, if we add (BF) to S_5Q and BQ to obtain S_5Q(BF) and BQ(BF), we have

$$\Sigma \vdash_{S_5Q(BF)} A \quad \text{iff} \quad \Sigma \vdash_{S_5Q} A.$$
$$\Sigma \vdash_{BQ(BF)} A \quad \text{iff} \quad \Sigma \vdash_{BQ} A.$$

The definitions of formal deducibility, formal provability, consistency, and maximal consistency with respect to the various systems are omitted.

Soundness and completeness of the various systems of modal first-order logic (without equality) will be studied in the following two sections. We may state in advance:

[1] TQ and S_4Q are sound and complete with respect to the semantics formulated in Definition 9.2.1.

[2] TQ(BF) and S_4Q(BF) are sound and complete with respect to the semantics formulated in Definition 9.2.4.

[3] S_5Q and BQ (equivalently S_5Q(BF) and BQ(BF)) are sound and complete with respect to the semantics formulated in both Definitions 9.2.1 and 9.2.4.

The theorems of replaceability of (both logically and syntactically) equivalent formulars hold in modal first-order logic as well.

Exercises 9.3.

9.3.1. Prove $L\forall xA(x) \vdash_{TQ} \forall xLA(x)$ (the converse of (BF)).

9.3.2. Prove $M\exists xA(x) \dashv\vdash_{TQ(BF)} \exists xMA(x)$
(use $L\forall xA(x) \dashv\vdash_{TQ(BF)} \forall xLA(x)$).

9.3.3. Prove $M\forall xA(x) \vdash_{TQ} \forall xMA(x)$.

9.3.4. Prove $\exists xLA(x) \vdash_{TQ} L\exists xA(x)$.

9.4. SOUNDNESS

Theorem 9.4.1. (***Soundness of TQ, S_4Q, S_5Q, BQ***)
Suppose $\Sigma \subseteq Form(\mathcal{L}^m)$ and $A \in Form(\mathcal{L}^m)$. Then

[1] If $\Sigma \vdash_{TQ} A$, then $\Sigma \models_{TQ} A$.

[2] If A is TQ-provable, then A is TQ-valid.

[3] If Σ is TQ-satisfiable, then Σ is TQ-consistent.

with respect to the semantics formulated in Definition 9.2.1. Similarly for S_4Q, S_5Q, and BQ.

Theorem 9.4.2. (***Soundness of TQ(BF), S_4Q(BF), S_5Q, BQ***)
Suppose $\Sigma \subseteq Form(\mathcal{L}^m)$ and $A \in Form(\mathcal{L}^m)$. Then

[1] If $\Sigma \vdash_{TQ(BF)} A$, then $\Sigma \models_{TQ(BF)} A$.

[2] If A is TQ(BF)-provable, then A is TQ(BF)-valid.

[3] If Σ is TQ(BF)-satisfiable, then Σ is TQ(BF)-consistent. with respect to the semantics formulated in Definition 9.2.4. Similarly for S_4Q(BF), S_5Q(BF) (which is equivalent to S_5Q), and BQ(BF) (which is equivalent to BQ).

The proof of the above theorems is left to the reader.

9.5. COMPLETENESS

Firstly we shall consider the completeness of the modal systems without the Barcon formula (with respect to the semantics formulated in Definition 9.2.1). We begin with TQ.

The function symbols will be omitted for simplicity.

As in the case of constructive logic (see Section 7.5), we suppose $\mathcal{D}_0$, $\mathcal{D}_1$, $\mathcal{D}_2$, ... are countable sets of new free variable symbols and $\mathcal{L}^m$, $\mathcal{D}_0$, $\mathcal{D}_1$, $\mathcal{D}_2$, ... are pairwise disjoint. Let

$$\mathcal{L}_0^m = \mathcal{L}^m,$$
$$\mathcal{L}_{n+1}^m = \mathcal{L}_n^m \cup \mathcal{D}_n \ (n \geq 0),$$
$$\mathcal{D} = \bigcup_{n \in N} \mathcal{D}_n,$$
$$\mathcal{L}^{m\prime} = \mathcal{L}^m \cup \mathcal{D}.$$

Then $Term(\mathcal{L}_n^m)$ and $Term(\mathcal{L}^{m\prime})$ are sets of terms of $\mathcal{L}_n^m$ and $\mathcal{L}^{m\prime}$; $Form(\mathcal{L}_n^m)$ and $Form(\mathcal{L}^{m\prime})$ are sets of formulas of $\mathcal{L}_n^m$ and $\mathcal{L}^{m\prime}$.

Suppose $\Sigma \subseteq Form(\mathcal{L}^m)$ is TQ-consistent. As in the case of T, we will construct $\Delta = \{\Sigma_1^*, \ldots, \Sigma_i^*, \ldots\}$, the elements of Δ being TQ-maximal consistent sets. The procedure will be described in detail.

Σ_1^* is constructed as follows. First, for each existential formula $\exists x A(x) \in Form(\mathcal{L}_1^m)$, we add successively $\exists x A(x) \to A(d)$ to Σ, d at each stage being some new symbol taken from $\mathcal{D}_1$, which has not yet occurred in Σ, nor in any previously added formula, nor in this $\exists x A(x)$ itself. Thus Σ is extended to some TQ-consistent set $\Sigma_1^\circ \subseteq Form(\mathcal{L}_1^m)$ such that for each $\exists x A(x) \in Form(\mathcal{L}_1^m)$, there is some $d \in \mathcal{D}_1$ such that $\exists x A(x) \to A(d) \in \Sigma_1^\circ$. Then we extend Σ_1° to some TQ-maximal consistent set $\Sigma_1^* \subseteq Form(\mathcal{L}_1^m)$ such that Σ_1^* has the E-property (see Definition 5.4.1). This is done by adding successively in the standard way, all those members of $Form(\mathcal{L}_1^m)$ which do not cause TQ-inconsistency. (The proof is left to the reader.)

It needs to be pointed out that Σ_1^* is here TQ-maximal consistent with respect to $Form(\mathcal{L}_1^m)$, that is, for each $A \in Form(\mathcal{L}_1^m)$ such that $A \notin \Sigma_1^*$,

1) $\Sigma_1^* \cup \{A\}$ is TQ-inconsistent.

We note that, if $A \notin Form(\mathcal{L}_1^m)$, we cannot conclude 1) from $A \notin \Sigma_1^*$.

Having obtained Σ_1^*, we then construct, for each $\Sigma_i^* \in \Delta$ (including Σ_1^* itself), a series of TQ-maximal consistent sets subordinate to Σ_i^* (each of which corresponds to some $MB \in \Sigma_i^*$ and is written as Σ_j^*) as follows.

First we take any MB $\in \Sigma_i^*$. Suppose Σ_j^* corresponding to MB is the k_jth set to be constructed in Δ. Let

$$\Sigma_j = \{\mathrm{B}\} \cup \{\mathrm{C} \mid \mathrm{LC} \in \Sigma_i^*\}.$$

As shown in Section 8.5, Σ_j is TQ-consistent. For each existential formula $\exists \mathrm{xA(x)} \in Form(\mathcal{L}_{k_j}^m)$, we add successively $\exists \mathrm{xA(x)} \to \mathrm{A(d)}$ to Σ_j and obtain Σ_j°, d at each stage being some new symbol taken from $\mathcal{D}_{k_j}$ which has not yet occurred before. (This can be done because, firstly, Σ_i^* precedes Σ_j^* in Δ. Hence assuming Σ_i^* is the k_ith set in Δ, we have $k_i < k_j$ and then $\Sigma_i^* \subseteq Form(\mathcal{L}_{k_i}^m)$, $\Sigma_j \subseteq Form(\mathcal{L}_{k_i}^m)$. Secondly, d at each stage is taken from $\mathcal{D}_{k_j}$, hence d does not occur in Σ_j.) Then Σ_j is extended to some TQ-consistent set $\Sigma_j^\circ \subseteq Form(\mathcal{L}_{k_j}^m)$ such that for each $\exists \mathrm{xA(x)} \in Form(\mathcal{L}_{k_j}^m)$, there is some $\mathrm{d} \in \mathcal{D}_{k_j}$ such that $\exists \mathrm{xA(x)} \to \mathrm{A(d)} \in \Sigma_j^\circ$. Next we extend Σ_j° to some TQ-maximal consistent set $\Sigma_j^* \subseteq Form(\mathcal{L}_{k_j}^m)$ such that Σ_j^* has the E-property. This is done by adding successively in the standard way, all those members of $Form(\mathcal{L}_{k_j}^m)$ which do not cause TQ-inconsistency. (The proof is left to the reader.) Then we have $\Sigma_i^* sub \Sigma_j^*$. We note that Σ_j^* is here TQ-maximal consistent with respect to $Form(\mathcal{L}_{k_j}^m)$.

To sum up: we have constructed $\Delta = \{\Sigma_1^*, \ldots, \Sigma_i^*, \ldots\}$; for each $\Sigma_i^* \in \Delta$, if we suppose Σ_i^* is the k_ith set, then Σ_i^* is TQ-maximal consistent with respect to $Form(\mathcal{L}_{k_i}^m)$. Now we will construct a valuation v_{k_i} for $\mathcal{L}_{k_i}^m$ over domain $D(v_{k_i})$ such that

$$\begin{aligned}
&D(v_{k_i}) = \{\mathrm{t}' \mid \mathrm{t} \in Term(\mathcal{L}_{k_i}^m)\},\\
&\mathrm{a}^{v_{k_i}} = \mathrm{a}',\\
&\mathrm{u}^{v_{k_i}} = \mathrm{u}',\\
&\mathrm{F}^{v_{k_i}} \subseteq D(v_{k_i})^m \quad \text{such that for any} \quad \mathrm{t}_1', \ldots, \mathrm{t}_m' \in D(v_{k_i}),\\
&\qquad\qquad \langle \mathrm{t}_1', \ldots, \mathrm{t}_m' \rangle \in \mathrm{F}^{v_{k_i}} \quad \text{iff} \quad \mathrm{F}(\mathrm{t}_1, \ldots, \mathrm{t}_m) \in \Sigma_i^*\\
&\qquad\qquad (\text{that is}, \mathrm{F}(\mathrm{t}_1, \ldots, \mathrm{t}_m)^{v_{k_i}} = 1 \quad \text{iff}\\
&\qquad\qquad \mathrm{F}(\mathrm{t}_1, \ldots, \mathrm{t}_m) \in \Sigma_i^*),
\end{aligned}$$

where a, u, F are any individual symbol, free variable symbol, m-ary relation symbol of $\mathcal{L}_{k_i}^m$.

Suppose $K = \{v_{k_i} \mid \Sigma_i^* \in \Delta\}$ and R is a binary relation on K such that $v_{k_i} R v_{k_j}$ iff $\Sigma_i^* = \Sigma_j^*$ or $\Sigma_i^* sub \Sigma_j^*$ ($v_{k_i}, v_{k_j} \in K$). Then R is reflexive.

Suppose $v_{k_i} R v_{k_j}$. If $\Sigma_i^* = \Sigma_j^*$, then $k_i = k_j$, and accordingly v_{k_i} is identical with v_{k_j} and $D(v_{k_i}) = D(v_{k_j})$. If $\Sigma_i^* sub \Sigma_j^*$, then $k_i < k_j$, and accordingly $\mathcal{L}^m_{k_i} \subseteq \mathcal{L}^m_{k_j}$ and $D(v_{k_i}) \subseteq D(v_{k_j})$. Hence K (with its elements) and R satisfy the conditions stated in Definition 9.2.1.

Lemma 9.5.1.

Suppose K and R are given as above. For every $\mathrm{A} \in Form(\mathcal{L}^{m\prime})$ and every $v_{k_i} \in K$, $\mathrm{A}^{v_{k_i}} = 1$ iff $\mathrm{A} \in \Sigma^*_i$.

Proof. By induction on the structure of A. We choose to prove for the cases of $\mathrm{A} = \neg\mathrm{B}$, LB, $\exists\mathrm{xB(x)}$. The remaining cases are left to the reader.

Case of $\mathrm{A} = \neg\mathrm{B}$. First, we prove $(\neg\mathrm{B})^{v_{k_i}} = 1 \Longrightarrow \neg\mathrm{B} \in \Sigma^*_i$. Suppose $(\neg\mathrm{B})^{v_{k_i}} = 1$. Then $(\neg\mathrm{B})^{v_{k_i}}$ and $\mathrm{B}^{v_{k_i}}$ are not undefined, and $\mathrm{B} \in Form(\mathcal{L}^m_{k_i})$. We have:

(1) $\mathrm{B}^{v_{k_i}} = 0$.
(2) $\mathrm{B} \notin \Sigma^*_i$ (by (1), ind hyp).
(3) $\neg\mathrm{B} \in \Sigma^*_i$ (by (2), Lem 5.3.3, TQ-max consis of Σ^*_i with respect to $Form(\mathcal{L}^m_{k_i})$, $\mathrm{B} \in Form(\mathcal{L}^m_{k_i})$).

For the converse, suppose $\neg\mathrm{B} \in \Sigma^*_i$. Then $\mathrm{B} \in Form(\mathcal{L}^m_{k_i})$ and accordingly $\mathrm{B}^{v_{k_i}}$ is not undefined. We have:

(4) $\mathrm{B} \notin \Sigma^*_i$.
(5) $\mathrm{B}^{v_{k_i}} = 0$ (by (4), ind hyp, $\mathrm{B}^{v_{k_i}}$ is not undefined).
(6) $(\neg\mathrm{B})^{v_{k_i}} = 1$.

Case of A = LB. First we prove $(\mathrm{LB})^{v_{k_i}} = 1 \Longrightarrow \mathrm{LB} \in \Sigma^*_i$. Suppose $(\mathrm{LB})^{v_{k_i}} = 1$. Then $(\mathrm{LB})^{v_{k_i}}$ is not undefined, and accordingly $\mathrm{LB} \in Form(\mathcal{L}^m_{k_i})$, $\mathrm{M}\neg\mathrm{B} \in Form(\mathcal{L}^m_{k_i})$. Suppose $\mathrm{LB} \notin \Sigma^*_i$. We have:

(7) For every $v_{k_j} \in K$ such that $v_{k_i} R v_{k_j}$, $\mathrm{B}^{v_{k_j}} = 1$ (by $(\mathrm{LB})^{v_{k_i}} = 1$).
(8) For every $\Sigma^*_j \in \Delta$ such that $\Sigma^*_i = \Sigma^*_j$ or $\Sigma^*_i sub \Sigma^*_j$, $\mathrm{B} \in \Sigma^*_j$ (by (7), ind hyp).
(9) $\neg\mathrm{LB} \in \Sigma^*_i$ (by $\mathrm{LB} \notin \Sigma^*_i$, Lem 5.3.3, TQ-max consis of Σ^*_i with respect to $Form(\mathcal{L}^m_{k_i})$, $\mathrm{LB} \in Form(\mathcal{L}^m_{k_i})$).
(10) $\mathrm{M}\neg\mathrm{B} \in \Sigma^*_i$ (by (9), $\mathrm{M}\neg\mathrm{B} \in Form(\mathcal{L}^m_{k_i})$, TQ-max consis of Σ^*_i with respect to $Form(\mathcal{L}^m_{k_i})$, $\neg\mathrm{LB} \vdash\!\dashv_{\mathrm{TQ}} \mathrm{M}\neg\mathrm{B}$).
(11) There is some $\Sigma^*_j \in \Delta$ such that $\Sigma^*_i sub \Sigma^*_j$ and $\neg\mathrm{B} \in \Sigma^*_j$, hence $\mathrm{B} \notin \Sigma^*_j$ (by (10), construction of Σ^*_j).

Since (11) contradicts (8), we obtain $LB \in \Sigma_i^*$. For the converse, suppose $LB \in \Sigma_i^*$. Then $B \in Form(\mathcal{L}_{k_i}^m)$. We have:

(12) $B \in \Sigma_i^*$ (by $LB \in \Sigma_i^*$, $B \in Form(\mathcal{L}_{k_i}^m)$, TQ-max consis of Σ_i^* with respect to $Form(\mathcal{L}_{k_i}^m)$).

(13) For every $\Sigma_j^* \in \Delta$ such that $\Sigma_i^* sub \Sigma_j^*$, $B \in \Sigma_j^*$ (by $LB \in \Sigma_i^*$, construction of Σ_j^*).

(14) For every $\Sigma_j^* \in \Delta$ such that $\Sigma_i^* = \Sigma_j^*$ or $\Sigma_i^* sub \Sigma_j^*$, $B \in \Sigma_j^*$ (by (12), (13)).

(15) For every $v_{k_j} \in K$ such that $v_{k_i} R v_{k_j}$, $B^{v_{k_j}} = 1$ (by (14), ind hyp).

(16) $(LB)^{v_{k_i}} = 1$ (by (15)).

Case of $A = \exists x B(x)$. First, we prove $\exists x B(x)^{v_{k_i}} = 1 \Longrightarrow \exists x B(x) \in \Sigma_i^*$. Suppose $\exists x B(x)^{v_{k_i}} = 1$. Then $\exists x B(x)^{v_{k_i}}$ is not undefined, and $\exists x B(x) \in Form(\mathcal{L}_{k_i}^m)$. We have:

(17) There is some $t' \in D(v_{k_i})$ such that $B(u)^{v_{k_i}(u/t')} = 1$, u being free variable symbol in $\mathcal{L}_{k_i}^m$, not occurring in $B(x)$ (by $\exists x B(x)^{v_{k_i}} = 1$).

(18) $B(t)^{v_{k_i}} = B(u)^{v_{k_i}(u/t')} = 1$ (by (17), $t^{v_{k_i}} = t'$).

(19) $B(t) \in \Sigma_i^*$ (by (18), ind hyp).

(20) $\exists x B(x) \in \Sigma_i^*$ (by (19), TQ-max consis of Σ_i^* with respect to $Form(\mathcal{L}_{k_i}^m)$, $\exists x B(x) \in Form(\mathcal{L}_{k_i}^m)$).

For the converse, suppose $\exists x B(x) \in \Sigma_i^*$. We have:

(21) There is some $d \in \mathcal{D}_{k_i}$ such that $B(d) \in \Sigma_i^*$ (by E-property of Σ_i^*).

(22) $B(d)^{v_{k_i}} = 1$ (by (21), ind hyp).

(23) $\exists x B(x)^{v_{k_i}} = 1$ (by (22)). □

Remarks

(1) In the above proof, the condition $B \in Form(\mathcal{L}_{k_i}^m)$, for instance, is necessary for establishing (3). If $B \notin Form(\mathcal{L}_{k_i}^m)$, we cannot derive $\neg B \in \Sigma_i^*$ from $B \notin \Sigma_i^*$ because Σ_i^* is TQ-maximal consistent with respect to $Form(\mathcal{L}_{k_i}^m)$. Similarly, $\exists x B(x) \in Form(\mathcal{L}_{k_i}^m)$ is necessary for establishing (20). (Refer to the note after 1).)

(2) Lemma 9.5.1 holds for S_4Q, S_5Q, and BQ as well. In the formulation and proof of this lemma for these systems, we will make the same modifications as for the corresponding systems of modal propositional logic, that

is, we will stipulate for S_4Q that $v_{k_i} R v_{k_j}$ iff $\Sigma_i^* = \Sigma_j^*$ or $\Sigma_i^* sub_n \Sigma_j^*$ for some $n \geq 1$, for S_5Q that $v_{k_i} R v_{k_j}$ iff $\Sigma_i^* = \Sigma_j^*$ or $\Sigma_i^* sub_n \Sigma_j^*$ for some $n \geq 1$ or $\Sigma_j^* sub \Sigma_i^*$, and for BQ that $v_{k_i} R v_{k_j}$ iff $\Sigma_i^* = \Sigma_j^*$ or $\Sigma_i^* sub \Sigma_j^*$ or $\Sigma_j^* sub \Sigma_i^*$. Then what needs to be added to the above proof of Lemma 9.5.1 for TQ is the same as that for the systems S_4, S_5, and B.

Theorem 9.5.2. (***Completeness of TQ, S_4Q, S_5Q, BQ***)
Suppose $\Sigma \subseteq Form(\mathcal{L}^m)$ and $A \in Form(\mathcal{L}^m)$. Then

[1] If Σ is TQ-consistent, then Σ is TQ-satisfiable.
[2] If $\Sigma \models_{TQ} A$, then $\Sigma \vdash_{TQ} A$.
[3] If A is TQ-valid, then A is TQ-provable.

with respect to the semantics formulated in Definition 9.2.1. Similarly for S_4Q, S_5Q, and BQ. □

Now we turn to the completeness of those systems with the Barcan formula.

The treatment for TQ(BF) is analogous to that for TQ mentioned above. Suppose $\Sigma \subseteq Form(\mathcal{L}^m)$ is TQ(BF)-consistent. Starting from Σ, we construct $\Delta = \{\Sigma_1^*, \ldots, \Sigma_i^*, \ldots\}$. Then for each $\Sigma_i^* \in \Delta$, suppose Σ_i^* is the k_ith set in Δ. We construct a valuation v_{k_i} for $\mathcal{L}_{k_i}^m$ over domain D such that

$$
\begin{aligned}
& D = \{t' \mid t \in Term(\mathcal{L}^{m\prime})\}; \\
& a^{v_{k_i}} = a'; \\
& u^{v_{k_i}} = u' \quad \text{(u being any free variable symbol in } \mathcal{L}^{m\prime}); \\
& F^{v_{k_i}} \subseteq D^m \quad \text{such that for any } t_1', \ldots, t_m' \in D, \\
& \qquad \langle t_1', \ldots, t_m' \rangle \in F^{v_{k_i}} \quad \text{iff} \quad F(t_1, \ldots, t_m) \in \Sigma_i^*.
\end{aligned}
$$

Let $K = \{v_{k_i} \mid \Sigma_i^* \in \Delta\}$. We note that the single domain D is available for each $v_{k_i} \in K$.

Similarly for S_4Q(BF), S_5Q(BF), and BQ(BF).

Then we can establish a lemma for these modal systems which is analogous to Lemma 9.5.1. The details are left to the reader.

Theorem 9.5.3. (***Completeness of TQ(BF), S_4Q(BF), S_5Q, BQ***)
Suppose $\Sigma \subseteq Form(\mathcal{L}^m)$ and $A \in Form(\mathcal{L}^m)$. Then

[1] If Σ is TQ(BF)-consistent, then Σ is TQ(BF)-satisfiable.

[2] If $\Sigma \models_{TQ(BF)} A$, then $\Sigma \vdash_{TQ(BF)} A$.

[3] If A is TQ(BF)-valid, then A is TQ(BF)-provable.

with respect to the semantics formulated in Definition 9.2.4. Similarly for $S_4Q(BF)$, $S_5Q(BF)$ (which is equivalent to S_5Q), and BQ(BF)(which is equivalent to BQ). □

9.6. EQUALITY

The usual interpretation of the equality symbol is $(t_1 \approx t_2)^v = 1$ iff $t_1^v = t_2^v$, where v is any valuation. For modal systems, however, we are faced with two kinds of semantics concerning equality.

Suppose K is any set of valuations and R is any binary relation on K, which may meet the requirements of any one of the systems TQ, S_4Q, S_5Q, and BQ. Suppose $v_i \in K$ and $(t_1 \approx t_2)^{v_i} = 1$, that is, $t_1^{v_i} = t_2^{v_i}$. According to the semantics of one kind, it is stipulated that for any $v_j \in K$ such that $v_i R v_j$, we have $t_1^{v_j} = t_2^{v_j}$ and hence $(t_1 \approx t_2)^{v_j} = 1$. In the case of $(t_1 \approx t_2)^{v_i} = 0$ (that is, $t_1^{v_i} \neq t_2^{v_i}$), it is stipulated that $t_1^{v_j} \neq t_2^{v_j}$ and hence $(t_1 \approx t_2)^{v_j} = 0$. Such an interpretation means that equal objects are necessarily equal and unequal objects are necessarily unequal. (In other words, it means that true propositions of equality are necessarily true, and false propositions of equality are necessarily false.) Accordingly we have:

1) $$t_1 \approx t_2 \models L(t_1 \approx t_2).$$

2) $$\neg(t_1 \approx t_2) \models L\neg(t_1 \approx t_2).$$

According to another kind of semantics, we can have the same individual assigned by v_i to t_1 and t_2 and hence $(t_1 \approx t_2)^{v_i} = 1$, but different individuals assigned by v_j to t_1 and t_2 and hence $(t_1 \approx t_2)^{v_j} = 0$. We can also have different individuals assigned by v_i to t_1 and t_2 and hence $(t_1 \approx t_2)^{v_i} = 0$, but the same individual assigned by v_j to t_1 and t_2 and hence $(t_1 \approx t_2)^{v_j} = 1$. Therefore, $(t_1 \approx t_2)^{v_i} = 1$ does not imply $(t_1 \approx t_2)^{v_j} = 1$, nor does $(t_1 \approx t_2)^{v_i} = 0$ imply $(t_1 \approx t_2)^{v_j} = 0$. Thus 1) and 2) do not hold.

Now we turn to formal deducibility. The rules of formal deduction for equality corresponding to the semantics of the first kind are exactly the same as in the case of classical first-order logic. They are:

($\approx -$) If $\Sigma \vdash A(t_1)$,
$\Sigma \vdash t_1 \approx t_2$,
then $\Sigma \vdash A(t_2)$, where $A(t_2)$ results from $A(t_1)$ by replacing some (not necessarily all) occurrences of t_1 in $A(t_1)$ by t_2.

($\approx +$) $\emptyset \vdash u \approx u$.

Adopting these rules we can derive in TQ (and accordingly in S_4Q, S_5Q, and BQ)

3)
$$t_1 \approx t_2 \vdash L(t_1 \approx t_2)$$

and derive in S_5Q and BQ

4)
$$\neg(t_1 \approx t_2) \vdash L\neg(t_1 \approx t_2).$$

The proof of 3) is as follows:

(1) $L(t_1 \approx t_1), t_1 \approx t_2 \vdash L(t_1 \approx t_2)$.
(2) $\emptyset \vdash t_1 \approx t_1$.
(3) $\emptyset \vdash L(t_1 \approx t_1)$ (by (L+), (2)).
(4) $t_1 \approx t_2 \vdash L(t_1 \approx t_1)$.
(5) $t_1 \approx t_2 \vdash t_1 \approx t_2$.
(6) $t_1 \approx t_2 \vdash L(t_1 \approx t_2)$ (by (4), (5), (1)).

The proof of 4) is as follows:

(1) $t_1 \approx t_2 \vdash L(t_1 \approx t_2)$ (by 3)).
(2) $t_1 \approx t_2 \vdash \neg M\neg(t_1 \approx t_2)$ (by (1)).
(3) $M\neg(t_1 \approx t_2) \vdash \neg(t_1 \approx t_2)$ (by (2)).
(4) $\neg(t_1 \approx t_2) \vdash L\neg(t_1 \approx t_2)$ (by Thm 8.3.8 [3], (3)).

3) and 4) may seem unacceptable intuitively. Hence weaker rules of formal deduction have been proposed in order that 3) and 4) cannot be derived. The rule ($\approx -$) may be replaced by a weaker rule

($\approx -'$) If $\Sigma \vdash A(t_1)$,
$\Sigma \vdash t_1 \approx t_2$,
then $\Sigma \vdash A(t_2)$, where $A(t_2)$ results from $A(t_1)$ by replacing some (not necessarily all) occurrences of t_1 not in the scope of any modal symbol by t_2.

Then the rules ($\approx -'$) and ($\approx +$) correspond to the second kind of semantics.

According to these two kinds of semantics and the rules of formal deduction corresponding to them, we can establish the soundness and completeness of various systems of modal first-order logic with equality. Soundness is proved by induction on the structure of $\Sigma \vdash A$. Completeness is established with the aid of the completeness of modal systems without equality. This is analogous to the case of classical first-order logic (see Chapter 5). The details are left to the reader.

The axiomatic deduction systems of various modal first-order logical systems (with or without equality) are obtained by adding axioms and rule of inference about modal symbols to the axiomatic deduction system of classical first-order logic, just as the axiomatic deduction system of T, S_4, S_5, and B are obtained from that of classical propositional logic (see Section 8.3).

APPENDIX

(a simple form of formal proof in natural deduction)

In this appendix we shall introduce a simple and clear form to facilitate the writing and reading of formal proofs in natural deduction.

In the form to be introduced, one formula is written on each line:

$$1)\qquad \left\{\begin{array}{llll} A_1 & & & \\ & A_2 & & \\ & & A_3 & \\ & & & A_4 \\ & & & B_1 \\ & & & B_2 \\ & & & B_3 \end{array}\right.$$

In this diagram A_1, A_2, A_3, A_4 are the premises. They are written in such a way that A_1 is the first premise and is written in the leftmost position. A_2, the second premise, is written on the right of A_1 (that is, the first symbol of A_2 is written on the right of that of A_1), and A_3 is written on the right of A_2, etc.

B_1 is not written on the right of A_4, but under it (that is, the first symbol of B_1 is aligned with that of A_4). Then B_1 is not a premise, but a conclusion. Similarly for B_2 and B_3. Therefore a formula in such a diagram is a premise iff it is written on the right of the formula immediately preceeding it.

Each formula which is a conclusion is intended to express a scheme of formal deducibility. The conclusion of the scheme expressed is the formula itself, and the premises of the scheme include the topmost formula over the conclusion and all those premises to the left of this topmost formula. In 1), for instance, the conclusions B_1, B_2, B_3 express, respectively,

$$A_1, A_2, A_3, A_4 \vdash B_1,$$
$$A_1, A_2, A_3, A_4 \vdash B_2,$$
$$A_1, A_2, A_3, A_4 \vdash B_3.$$

We note that B_1 is not included in the premises for B_2 or B_3, because B_1 is not the topmost formula over B_2 or B_3. B_1 is not a premise. Similarly for B_2.

Then the formal proof:

$$2)\quad \begin{cases} (1) & A \to B, B \to C, A \vdash A \to B \\ (2) & A \to B, B \to C, A \vdash A \\ (3) & A \to B, B \to C, A \vdash B \\ (4) & A \to B, B \to C, A \vdash B \to C \\ (5) & A \to B, B \to C, A \vdash C \\ (6) & A \to B, B \to C \vdash A \to C \end{cases}$$

can be written in the following form:

$$3)\quad \begin{cases} (1) & A \to B \\ (2) & \qquad B \to C \\ (3) & \qquad\qquad A \\ (4) & \qquad\qquad A \to B \quad (\text{by } (\epsilon)) \\ (5) & \qquad\qquad A \quad (\text{by } (\epsilon)) \\ (6) & \qquad\qquad B \quad (\text{by } (\to -), (4), (5)) \\ (7) & \qquad\qquad B \to C \quad (\text{by } (\epsilon)) \\ (8) & \qquad\qquad C \quad (\text{by } (\to -), (7), (6)) \\ (9) & \qquad A \to C \quad (\text{by } (\to +), (8)) \end{cases}$$

in which the formulas (conclusions) in (4)–(9) of 3) express, respectively, the schemes in (1)–(6) of 2). We note that $A \to C$ in (9) of 3) is written under $B \to C$ in (2). This means that $A \to C$ is not a premise, but a conclusion, and that the premises for $A \to C$ include $A \to B$ and $B \to C$. Hence $A \to C$ expresses the scheme in (6) of 2).

Obviously 3) is simpler and clearer than 2). But 3) can be further simplified, because the steps (4), (5), and (7) (using (ϵ)) occur repeatedly and may be deleted. 3) may be simplified as:

$$4)\quad \left\{\begin{array}{llll} (1) & A \to B & & \\ (2) & \quad B \to C & & \\ (3) & \qquad A & & \\ (4) & \qquad B & & (\text{by } (\to -), (1), (3)) \\ (5) & \qquad C & & (\text{by } (\to -), (2), (4)) \\ (6) & \quad A \to C & & (\text{by } (\to +), (5)) \end{array}\right.$$

We note that $A \to B$ in (1) of 3) is a premise, while $A \to B$ in (4) of 3) is a conclusion. Similarly for $B \to C$ and A in 3). But in 4), $A \to B$, $B \to C$, and A occur not only as premises but as conclusions as well, because some steps are deleted.

There is another advantage of this new form, which we shall describe. Suppose we are to prove $A \to B, B \to C \vdash A \to C$. We may first write the proof as

$$5)\quad \left\{\begin{array}{l} A \to B \\ \quad B \to C \\ \quad\quad . \\ \quad\quad . \\ \quad\quad . \\ \quad A \to C \end{array}\right.$$

and then add A and C to 5) in the following way:

$$\begin{array}{l} A \to B \\ \quad B \to C \\ \qquad A \\ \qquad . \\ \qquad . \\ \qquad . \\ \qquad C \\ \quad A \to C \end{array}$$

Thus 5) can be obtained from it by applying ($\rightarrow$+). The blanks between A and C can be filled easily.

The following are some examples to show how this new form is used in proving schemes of formal deducibility.

Example
If Σ, A $\vdash$ B,
Σ, A $\vdash$ $\neg$B,
then $\Sigma \vdash \neg$A.

Proof.
(1) Σ
(2) $\neg\neg$A
(3) A (by $\neg\neg A \vdash A$, (2))
(4) B (by supposition, (1), (3))
(5) $\neg$B (the same as (4))
(6) $\neg$A (by ($\neg-$), (4), (5)) □

Example
$\neg(A \vee B) \vdash \neg A \wedge \neg B$

Proof.
(1) $\neg(A \vee B)$
(2) A
(3) $A \vee B$ (by ($\vee$+), (2))
(4) $\neg$A (by ($\neg$+), (3), (1))
(5) $\neg$B (analogous to (4))
(6) $\neg A \wedge \neg B$ (by ($\wedge$+), (4), (5)) □

Example
$\forall x A(x) \rightarrow B \vdash \exists x(A(x) \rightarrow B)$

Proof.
(1) $\forall x A(x) \rightarrow B$
(2) $\neg\exists x(A(x) \rightarrow B)$
(3) $\forall x \neg(A(x) \rightarrow B)$ (by Thm 3.5.3 [2], (2))
(4) $\neg(A(u) \rightarrow B)$ (by ($\forall-$), (3), u not in (1))
(5) A(u) (by Thm 2.6.7 [5], (4))
(6) $\neg$B (by Thm 2.6.7 [6], (4))

(7) $\forall x A(x)$ (by ($\forall+$), (5))
(8) B (by ($\rightarrow-$), (1), (7))
(9) $\exists x(A(x) \rightarrow B)$ (by ($\neg-$), (8), (6)) □

Example
$\forall x A(x) \vee \forall x B(x) \vdash \forall x(A(x) \vee B(x))$

Proof. We first prove

$$\forall x A(x) \vdash \forall x(A(x) \vee B(x))$$

as follows:

(1) $\forall x A(x)$
(2) $A(u)$ (by ($\forall-$), (1), u not in $A(x)$)
(3) $A(u) \vee B(u)$ (by ($\vee+$), (2))
(4) $\forall x(A(x) \vee B(x))$ (by ($\forall+$), (3))

Similarly for

$$\forall x B(x) \vdash \forall x(A(x) \vee B(x)).$$

Then the theorem is proved by ($\vee-$). □

Example
$\exists x(A(x) \rightarrow B) \vdash \forall x A(x) \rightarrow B$, x not occurring in B.

Proof. We first prove

$$A(u) \rightarrow B \vdash \forall x A(x) \rightarrow B$$

as follows:

(1) $A(u) \rightarrow B$
(2) $\forall x A(x)$
(3) $A(u)$ (by ($\forall-$),(2))
(4) B (by ($\rightarrow-$), (1), (3))
(5) $\forall x A(x) \rightarrow B$ (by ($\rightarrow+$),(4))

We may take u not occurring in (5). Then the theorem is proved by ($\exists-$).
□

BIBLIOGRAPHY

Chang, C. C., and H. J. Keisler

[1973] *Model Theory.* (North-Holland, Amsterdam.)

Church, A.

[1936] A note on the Entscheidungsproblem, *J. Symb. Logic* **1**, 40–41. (Reprinted with corrections in Davis [1965], 110–115.)

Davis, M. (ed.)

[1965] *The Undecidable. Basic Papers on Undecidable Propositions, Unsovable Problems, and Computable Functions.* (Raven Press, New York.)

Glivenko, V.

[1929] Sur quelques points de la logique de M. Brouwer, *Bull. Acad. Roy. Belg. Sci.* (5) **15**, 183–188.

Gödel, K.

[1930] Die Vollständigkeit der Axiome des logischen Funktionenkalküls, *Monatsh. Math. Phys.* **37**, 349–360. (English transl. in Van Heijenoort [1967], 582–591.)

Henkin, L.

[1949] The completeness of the first-order functional calculus, *J. Symb. Logic* **14**, 159–166.

Herbrand, J.

[1930] Recherches sur la théorie de la démonstration, *Trav. Soc. Sci. Lett. Varsovie*, Cl. III **33**, 33–160. (English transl. of Ch. 5 in Van Heijenoort [1967], 525–581.)

Hughes, G. E., and M. J. Cresswell

[1968] *An Introduction to Modal Logic.* (Methuen and Co. Ltd.)

Kripke, S. A.

[1965] Semantical analysis of intuitionistic logic I. *Formal Systems and Recursive Functions*, 92–130, eds. J. N. Crossley and M. A. E. Dummett. (North-Holland, Amsterdam.)

Löwenheim, L.

[1915] Über Möglichkeiten im Relativkalkül, *Math. Ann.* **76**, 447–470. (English transl. in Van Heijenoort [1967], 228–251.)

Manna, Z.

[1982] Verification of sequential programs: temporal axiomatization. *Theoretical Foundations of Programming Methodology*, 53–102, eds. M. Broy and G. Schmidt. (D. Reidel Publishing Company, Holland.)

Skolem, T.

[1920] Logisch-kombinatorische Untersuchungen über die Erfüllbarkeit oder Beweisbarkeit mathematischer Sätze nebst einem Theoreme über dichte Mengen I. *Skr. Norske Vid.-Akad. Kristiana Mat.-Naturv.* kl. (4). (English transl. of Sec. 1 in Van Heijenoort [1967], 252–263.)

Van Heijenoort, J. (ed.)

[1967] *From Frege to Gödel, a Source Book in Mathematical Logic 1879–1931.* (Harvard Univ. Press, Cambridge, Mass.)

LIST OF SYMBOLS

The numbers refer to the pages on which the symbol occurs (or its meaning is explained) for the first time.

$\in$ 5

$\notin$ 5

iff 6

$\subseteq$ 6

$\emptyset$ 6

$\{x| \_ x \_\}$ 7

$-$ 7

$\cup$ 7

$\cap$ 7

$\bigcup_{i\in I}$ 7

$\bigcap_{i\in I}$ 7

$\langle\alpha,\beta\rangle$ 7

$\langle\alpha_1,\ldots,\alpha_n\rangle$ 8

$S_1\times\ldots\times S_n$ 8

S^n 8

dom 9

ran 9

$f:S\to T$ 9

$f|S$ 9

$\sim$ 10, 137

$|S|$ 10

N 12

ind hyp 12

$\Longrightarrow$ 15

$\Longleftrightarrow$ 15

$\Longleftarrow$ 15

Def 15

Thm 15

Lem 15

Cor 15

$\Box$ 15

$\mathcal{L}^p$ 21

p q r 22

$\neg$ 22

$\wedge$ 22

$\vee$ 22

$\rightarrow$ 22

$\leftrightarrow$ 22

(22

) 22

Atom ($\mathcal{L}^p$) 23

* 23

Form ($\mathcal{L}^p$) 23

($\neg$A) 23

(A*B) 23

A B C 24

deg(A) 26

t 35

p^t 35

A^t 35

Σ 36

Σ^t 36

$\Sigma \models A$ 41

$\Sigma \not\models A$ 41

$A \models\mid B$ 41

$\vdash$ 46

(Ref) 46

($\wedge-$) 46

($\wedge+$) 47

($\vee-$) 47

($\vee+$) 47

($\rightarrow -$) 47

($\rightarrow +$) 47

($\leftrightarrow -$) 47

($\leftrightarrow +$) 47

(ϵ) 48

$\Sigma \vdash A$ 50

$\Sigma \nvdash A$ 50

(Tr) 53

($\neg+$) 55

$A \vdash\dashv B$ 56

$\dashv$ 56

$\mid$ 66

$\downarrow$ 66

$\mathcal{N}$ 70

$\mathcal{G}$ 70

$\mathcal{L}$ 74

a b c 74

F G H 74

$\approx$ 74

f g h 74

u v w 75

x y z 75

$\forall$ 75

$\exists$ 75

$\forall$x 75

$\exists$x 75

, 75

$\mathcal{L}(\mathcal{N})$ 76

$\mathcal{L}(\mathcal{G})$ 76

Term($\mathcal{L}$) 76

$f(t_1, \ldots, t_n)$ 76

t 77

$U(s_1, \ldots, s_n)$ 78

$U(V_1, \ldots, V_n)$ 78

Atom($\mathcal{L}$) 78

$F(t_1, \ldots, t_n)$ 78

$\approx (t_1, t_2)$ 78

$t_1 \approx t_2$ 78

$Form(\mathcal{L})$ 78

$\forall xA(x)$ 78

$\exists xA(x)$ 78

v 87

t^v 88

A^v 88

$v(u/\alpha)$ 88

Σ^v 91

$(\forall -)$ 98

$(\forall +)$ 98

$(\exists -)$ 98

$(\exists +)$ 98

$(\approx -)$ 98

$(\approx +)$ 98

$\forall x_1 \ldots x_n$ 99

$\exists x_1 \ldots x_n$ 99

$\forall x_1 \ldots x_n A(x_1, \ldots, x_n)$ 100

$\exists x_1 \ldots x_n A(x_1, \ldots, x_n)$ 100

Q 102

$\exists!!$ 103

$\exists!$ 103

(Ax1) 109

(Ax2) 109

(Ax3) 109

(Ax4) 109

(Ax5) 109

(Ax6) 109

(Ax7) 109

(Ax8) 109

(Ax9) 109

(Ax10) 110

(Ax11) 110

(Ax12) 110

(Ax13) 110

(Ax14) 110

(Ax15) 110

(Ax16) 110

(Ax17) 110

(Ax18) 110

(R1) 110

(R2) 110

$\Sigma \vdash A$ 110

consis 127

max consis 127

T 134

t' 134

$\bar{t}$ 137

$\overline{T}$ 138

H_A 151

H 151

K 160

$\models_C$ 163

$(\neg)$ 164

$\vdash_C$ 164

$\vdash_C$ 164

L 180

M 180

$\mathcal{L}^{pm}$ 180

$Atom(\mathcal{L}^{pm})$ 180

$Form(\mathcal{L}^{pm})$ 180

(LA) 180

S_5 183
T 183
S_4 183
B 183
$\models_{S_5}$ 184
$\models_T$ 184
$\models_{S_4}$ 184
$\models_B$ 184
(MA) 185
(L−) 186
(L+) 186
$(\rightarrow -(L))$ 186
(L+L) 186
(L+M) 186
(LM+) 186
$\vdash_T$ 186
$\vdash_{S_4}$ 186
$\vdash_{S_5}$ 186
$\vdash_B$ 186
Rep eq 188
Δ 196
sub 196
sub_n 198
$\mathcal{L}^m$ 205
$Term(\mathcal{L}^m)$ 205
$Atom(\mathcal{L}^m)$ 205
$Form(\mathcal{L}^m)$ 205
TQ 206
S_4Q 206
S_5Q 206
BQ 206
$\models_{TQ}$ 207
$\models_{S_4Q}$ 207
$\models_{S_5Q}$ 207
$\models_{BQ}$ 207
TQ(BF) 208
$S_4Q(BF)$ 208
$S_5Q(BF)$ 208
BQ(BF) 208
$\models_{TQ(BF)}$ 208
$\models_{S_4Q(BF)}$ 208
$\models_{S_5Q(BF)}$ 208
$\models_{BQ(BF)}$ 208
BF 208
(BF) 210
$\vdash_{TQ}$ 210
$\vdash_{S_4Q}$ 210
$\vdash_{S_5Q}$ 210
$\vdash_{BQ}$ 210
$\vdash_{TQ(BF)}$ 210
$\vdash_{S_4Q(BF)}$ 210
$\vdash_{S_5Q(BF)}$ 210
$\vdash_{BQ(BF)}$ 210
$(\approx -')$ 218

INDEX

The numbers refer to the pages on which the term (or the same term of different meanings) occurs for the first time.

addition of premise 46
adequate set of connectives 65
algorithm 31
antecedent 29
assignment 86
associative law 43
associativity
  $\wedge$- 56
  $\vee$- 57
  $\leftrightarrow$- 58
atom 23, 78
atomic formula 23
axiom 109
axiomatic deduction system 109

B-
  consistency 186
  maximal consistency 186
  provability 186
  satisfiability 184
  validity 184
  valuation 184
Barcan, Ruth C. 208

Barcan formula 208
basis of induction 12
bound variable 72
bound variable symbol 75
BQ–
  consistency 210
  maximal consistency 210
  provability 210
  satisfiability 207
  validity 207
  valuation 206
BQ(BF)–
  consistency 210
  maximal consistency 210
  provability 210
  satisfiability 208
  validity 208
  valuation 208
Brouwer, L. E. J. 184

C–
  consistency 164
  formally provable 164
  logical consequence 163
  maximal consistency 164
  satisfiability 163
  validity 163
cardinal 10
cardinal number 10
Cartesian product 8
Chang, C. C. 180
Church, A. 92
classical first-order logic 69
classical logic 17
classical propositional logic 17
clause 61
closed formula 80
closed term 77
commutative law 43

commutativity
  $\wedge$– 56
  $\vee$– 57
  $\leftrightarrow$– 58
comma 75
compactness 147
complement 7, 63
complementary 63
complete 118
Completeness Theorem 131, 132, 136, 139, 178, 198, 203, 216
compound proposition 18
conclusion 1
conjunct 29
conjunction 22
conjunction formula 29
conjunctive normal form 61
  full — 63
connective 18, 22
consequent 29
consistency 126
constructive 158
constructive valuation 160, 161
contradiction 37
Cresswell, M. J. 184
countable 11
countably infinite 11
course-of-values induction 13

Deduction theorem 111
definition by recursion 13
degree of complexity of A $\in Form(\mathcal{L}^p)$ 26
De Morgan's Law 57
difference 7
disjoint 7
disjunct 29
disjunction 22
disjunction formula 29
disjunction property 172
disjunctive normal form 61

full — 63
domain 70
domain of functions 9
dual 44, 97
duality 44, 97

E-property 133
element 5
elementary logic 69
elimination
$\neg$– 46
$\wedge$– 47
$\vee$– 47
$\rightarrow$– 46
$\leftrightarrow$– 47
$\forall$– 98
$\exists$– 98
$\approx$– 98
L– 186
$\rightarrow$-elimination in the scope of L 186
empty expression 22
empty set 6
enumerable 11
enumerably infinite 11
equality symbol 74
equipotent 10
equivalence 22, 29
equivalence class 10
equivalence formula 29
equivalence relation 10
equivalent formulas 41, 56, 93
exclusive or 19
existence property 133
existential formula 81
existential quantifier 71, 75
existential quantifier symbol 75
expression 22
$\exists$-free prenex normal form 149

falsehood 18
first-order language 74
first-order logic 69
formal deducibility 46, 50, 99
    closed under — 128
    scheme of — 48
    transitivity of — 53
formal deduction 46, 110
    rule of — 46
formal language 3
formal proof 50, 110
formally deducible 46, 50, 110
formally provable 46, 59, 110
formation rules of formulas 23, 78
formation rules of terms 76
formation sequence 25
formula 23, 78
free variable 72
free variable symbol 75
function 8
function symbol 74

Glivenko, V. 167
Gödel, K. 133
Gödel's completeness theorem 133
Gödel translation 168

Henkin, L. 133
Herbrand, J. 154
Herbrand theorem 153
Herbrand universe 151
Herbrand valuation 152
higher-order logic 74
Hughes, G. E. 184

if-then-else 66
implication 21, 29
implication formula 29
impossible 179

inclusive or 19, 22
independence 140
indirect proof 49
individual 70
individual symbol 74
induction hypothesis 12
induction on the structure of formal deducibility 51
induction on the structure of formulas 25, 80
induction on the structure of terms 77
induction proposition 12
induction step 12
induction variable 12
inductive definition 11
inductive proof 12
initial segment 23
injection 9
instance 153
interpretation 84
intersection 7
introduction
- $\neg$– 55
- $\wedge$– 47
- $\vee$– 47
- $\rightarrow$– 46
- $\leftrightarrow$– 47
- $\forall$– 98
- $\exists$– 98
- $\approx$– 98
- L– 186
- L–introduction to L 186
- L–introduction to M 186
- LM– 186

Keisler, H. J. 181
Kripke, S. A. 159, 185

law of excluded middle 57
law of non-contradiction 56
left parenthesis 22

Leibniz, G. W. 3, 181
length of an expression 22
Lindenbaum, A. 129
literal 61
logical consequence 93
logical symbol 75
logically equivalent formulas 93
Löwenheim, L. 149
Löwenheim–Skolem theorem 148, 149
  downward — 149
  upward — 149

Manna, Z. 185
mapping 9
matrix 106
maximal consistency 127
member 5
metalanguage 3
modal first-order language 205
modal first-order logic 205
modal logic 179
modal operator 179
modal propositional language 180
modal propositional logic 179
model 126

natural deduction 60
necessary 179
necessity 179
necessity symbol 180
negation 22, 29
negation formula 29
non-constructive 158
non-logical symbol 75

object language 3
one-one function 9
ordered n-tuple 8
ordered pair 7

possibility 179
possibility symbol 180
possible 179
Post, E. L. 127
predicate logic 69
prefix 106
premise 1
prenex normal form 106
proof by cases 49
proof by induction 12
proper initial segment 23
proper segment 23
proper subset 6
proper terminal segment 23
proposition 1
proposition function 71
proposition symbol 22
propositional language 21
propositional logic 17
punctuation 22, 75

quantification 72
quantified 72
quantifier 71, 75
quantifier symbol 75
quantifier with restricted range 73
quasi-formula 80

range of functions 9
reductio ad absurdum 54
reduction laws 191
reflexive relation 10
reflexivity 46
relation 8
relation symbol 74
relational calculus 69
replaceability of bound variable symbols 106
replaceability of equivalent formulas 44, 59, 97, 105, 170, 185, 188, 211
restricted functional calculus 69

restricted predicate calculus 69
restricted quantifier 73
restriction of a function 9
restriction of a relation 9
right parenthesis 22
rule of inference 110

S_4–
  consistency 186
  maximal consistency 186
  provability 186
  satisfiability 184
  validity 184
  valuation 184
S_5–
  consistency 186
  maximal consistency 186
  provability 186
  satisfiability 183
  validity 183
  valuation 183
satisfiability 36, 91
satisfiability in a domain 119
scope 29, 82
second-order logic 73
segment 23
semantics 3
sentence 80
set 5
Sheffer stroke 66
Skolem, T. 149
simple proposition 18
sound 117
Soundness Theorem 125, 126, 171, 192, 194, 211
S_4Q–
  consistency 210
  maximal consistency 210
  provability 210
  satisfiability 207

validity 207
valuation 206
S_5Q–
consistency 210
maximal consistency 210
provability 210
satisfiability 207
validity 207
valuation 206
$S_4Q(BF)$–
consistency 210
maximal consistency 210
provability 210
satisfiability 208
validity 208
valuation 208
$S_5Q(BF)$–
consistency 210
maximal consistency 210
provability 210
satisfiability 208
validity 208
valuation 208
strong consistency 172
structure 70
subordinate 196
subset 6
surjection 9
symmetric relation 10
syntactically equivalent formulas 56
syntax 3

T–
consistency 186
maximal consistency 186
provability 186
satisfiability 184
validity 184
valuation 184

tautological consequence 41
tautologically equivalent formulas 41
tautology 37
term 76
terminal segment 23
theory of quantification with equality 69
TQ–
  consistency 210
  maximal consistency 210
  provability 210
  satisfiability 207
  validity 207
  valuation 206
TQ(BF)-
  consistency 210
  maximal consistency 210
  provability 210
  satisfiability 208
  validity 208
  valuation 208
transitive relation 10
true 92
truth 18
truth function 22
truth table 34, 37
truth valuation 35
truth value 18
truth-functional 179

undefined 162
union 7
universal formula 81
universal quantifier 71, 75
universal quantifier symbol 75
universal validity 92

validity 91
validity in a domain 119
valuation 87

value 18
value of formulas 35, 88, 161, 162, 181, 182, 207, 208
value of terms 88
variable 70

well-formed formula 23
world 180